Ron Ron

CW01620103

AIRFIELD HEYDAY

Paul Berry

© 1989 Paul Berry

Published by Paul Berry
11 Hayfield Road, North Wootton
King's Lynn PE30 3RR

ISBN 0 9515257 0 0

By the same author:
The Cries of Ashes (Scree, 1979)
Legacies (Scree, 1981)
Earth Musk and Country Dark (Outposts, 1982)
A Bequest of Fire (Outposts, 1985)
Homages and Holiday Snaps (Envoi, 1989)

Produced by Jim Baldwin Publishing Fakenham, Norfolk
Printed in England

A Product of Fakenham

For mum and dad
who were part of the places
and knew the days and ways
remembered here.

Acknowledgements

Thanks are due to many kind people who willingly gave their time to share airfield memories. Regrettably it has not proved possible to identify by name all contributors in the pages which follow but under this heading I can acknowledge the help afforded me by Colin and Grace Muddimer, Mollie McKenzie, Don Jones, Mrs K J Lucas, Joan Horton, Linda Dobbing, Eileen Jackson, John C H Peard, David Williams and Alan Harrison.

Of the illustrations in these pages, many are published for the first time. They were often taken 'illegally' and processed with difficulty. Airfields were restricted areas where unofficial picture taking was banned and, at a time of shortages, films and photographic processing were luxuries not easily available. I am therefore especially grateful to Sam Hurry, Bernard Beaumont, Edith Kup and Bill Muddimer for loaning pictures. I am pleased to acknowledge the help of Tom Cushing who opened up his private museum and gave me access to a superb collection of airfield related photographs. Some other pictures did not reach the final selection process but to everyone who searched treasured albums and souvenir boxes I offer thanks.

The task of contacting people who would play a part in the chapters which follow would have been harder but for the help of magazines and newspapers who allowed me access to their columns to appeal for information. For bridge building in this way I am indebted to the editors of the *SUFFOLK FREE PRESS*, the *NORWICH MERCURY SERIES*, the *LINCOLNSHIRE ECHO*, *LINCOLNSHIRE LIFE*, the *YORKSHIRE GAZETTE AND HERALD SERIES*, and *AIRMAIL*, the magazine of the Royal Air Forces Association. For quotations from their pages I am grateful to the *HONINGTON HERALD* and the *DRIFFIELD LEADER*.

Two books never off my desk during the last 3 years have been *BOMBER COMMAND: THE AIR MINISTRY ACCOUNT OF BOMBER COMMAND'S OFFENSIVE AGAINST THE AXIS* (HMSO, 1941) and Michael J F Bowyer's volume on East Anglia in the *ACTION STATIONS* series published by Patrick Stephens Ltd. To their respective authors I acknowledge the stepping stones which their work has provided in my own research.

The job of realising *AIRFIELD HEYDAY* in book form was accepted by Jim Baldwin who brought to the project more than I could have expected from an editor and publishing consultant. For assistance and enthusiasm beyond the call of duty, my thanks Jim.

Finally, and with great pleasure, I must record the contribution made by my wife. She shared the moment when the book began and without demur accepted my absences as I travelled to interview informants or clossetted myself with paper and pen. Coffee and encouragement were always provided at just the right time and as the project neared completion she cast a keen eye over the manuscript to prevent fewer errors than might have been. Thank you, Tina.

Bibliography

Fields of Little America: Martin BOWMAN (Wensum Books, 1977)
Action Stations: Various Authors – nine volumes (Patrick Stephens Ltd, 1979 –1982)
Wings Over Eastern England: G M DIXON and J RIPPON (Minimax Books, 1980)
Cologne: The First 1000 Bomber Raid: Charles MESSENGER (Ian Allan, 1982)
Bomber Command: The Air Ministry Account of Bomber Command's Offensive Against the Axis September 1939 – July 1941 (HMSO, 1941)
100 Bomb Group Memorial Museum: guidebook – anonymous (100th Bombardment Group Association, c1986)
East Anglia at War: Derek E JOHNSON (Jarrold, 1978)
40 Years of RAF Sculthorpe: Jim BALDWIN (Jim Baldwin Publishing, 1989)

Introduction

My wife and I were driving along a familiar route, an open road without trees or hedges crossing the kind of East Anglian landscape where eyes can so easily tire of looking for new things. But this time it was different: noise and activity was coming from the space around us and its source owed nothing to agriculture which was the usual reason for bustle.

That day there were men with pneumatic drills, compressors, front loading excavators and dump trucks, up to their waists and tyre tops in waves of grass. This was no ordinary field. It had once been part of a wartime airfield. The runways had been broken up long since and the area returned to growing. Now only the taxiways and a few dispersals which had survived were being removed so that farming could take over totally.

We stopped the car and sat watching men breaking the concrete's surface with drills, saw the excavator and dumper ply fussily between them and the large dark machine with rotating drum and spinning flywheel. The rubble was fed in at one end and amid dust and the noise of stone being crushed coming from the very bowels of the machine, a pyramid of dust and small stones grew at the other end as small pieces of former runway fell from a conveyor belt.

It was impossible to watch and not imagine what the place would have been like nearly fifty years ago when the runways were intact and camouflaged buildings were scattered around the same fields among the trees. The sound would have been of heavy aircraft engines, the activity not of a group of workmen, but of hundreds of men and women all working to get aircraft airborne.

It was a contrast to the rural idyll surrounding us. The hangars had been removed long ago. Technical and domestic buildings had been demolished, those few remaining soon capitulated to briars and creeping ivy. Slowly the forces of man and nature were reclaiming the site and in a matter of months there would be nothing apart from memories to remind what had happened here.

Watching the breaking and trying to image how it was we became aware of how little we knew about what had gone on there. Military histories and squadron archives could tell us about the battles and missions but they glossed over the everyday details of what it was like to live and work on an airfield at a time when the country was at war. There was only one way to gain those kind of insights and that would involve talking to people who were there. And if the task was to be done it would need to be done soon for just as the airfields themselves were being lost, so the people who lived and worked on and around them were getting older. (Indeed, in the time between research and publication, John Wortley and Diarmud Collins have died.) And so the idea for this book evolved: to gain an insight into daily life on airfields through the experiences of people who lived and worked on or near them.

In helping recreate this impression of those times long gaps in memory have had to be bridged, and memory can play tricks. The people I have talked to have shared their memories in all honesty with no wish to deceive or sensationalise. Their stories are recounted here in that same spirit yet with the knowledge that, by the very nature of the task, some detail may have been lost, or an extra lustre added with time. No apology is offered since what follows is not intended as the definitive statement of airfield life. It is merely one attempt to recreate through personal memories and anecdotes an impression of what life was like on some stations. On others, with different Commanding Officers, it was probably different although there are threads common to all.

What follows may act as a springboard, reawakening other memories long dormant in the minds of ex-servicemen and civilians and this book may be the first of many. For those who lived near airfields and never knew what went on, apart from the regular taking off and landing of planes, these pages may provide a long overdue insight into the lives of their one time neighbours. And for those of a younger generation there may be answers to questions like whatever happened in that building, did people really live in those tin huts, and of course, what did you do in the war grandad?

Two years ago we stopped our car and watched an essential, and we once thought, enduring part of an airfield, being broken up. Soon on that particular site there would be nothing to remind succeeding generations how different those same fields once were; nothing to recall how a few hundred acres in the middle of nowhere suddenly became a community. While the old airfields are broken up and history is destroyed the words which follow may hopefully recreate and restore something of their essence long after they are lost to landscape and living memory.

PAUL BERRY
King's Lynn
July 1989

1 First Fields

Many of the airfields which would play a crucial role in the Second World War had beginnings dating back to the earliest days of military flying, when the Royal Air Force did not even exist. In 1912 when the first airbase was opened on the Suffolk coast at Orfordness, military flying was shared between the Army which was responsible for the Royal Flying Corps and the Navy which controlled the Royal Naval Air Service. The two senior services had been quick to realise the potential of flying machines and airships and some of the first bases for them were located along the eastern seaboard of England. There the land was flatter than elsewhere and well suited to flying operations. And, with the eventual outbreak of war, its proximity to Europe was to be of strategic importance.

Among the earliest airfields was one designated near the village of Narborough in Norfolk, acquired by the Navy for use as a base from which their frail biplanes could maintain coastal patrols and harry German Zeppelin airships during the First World War. Also at this time balloon reconnaisance on the battlefield was giving way to observers in flying machines with the result that Army chiefs were needing to build airfields where young pilots could train and receive basic instruction in flying techniques before being posted to France.

As World War One progressed German Zeppelins began achieving greater success at breaking through the Naval air defence and finding themselves over mainland Britain. Many cities and towns including Leeds, Manchester, Liverpool, Nottingham, Norwich, Ipswich and Dover were to suffer from the explosive power of bombs dropped from gondolas hanging under the huge belly of the airships. Nowhere seemed safe when the defences had been penetrated. There followed a review of provisions to protect civilians from long range air attack and so, along an area stretching from Hull in the north to London in the south, a Home Defence Network was formed. This consisted of airfields sited throughout the zone, each with its own flight of aircraft supported by servicing facilities. Fighter aircraft not considered suitable for front line service in France were brought back by the Royal Flying Corps and based at Home Defence Network Airfields which were sited near large towns. Alongside the operational fields were emergency landing sites for planes in trouble or needing a safe haven from suddenly changing weather patterns.

Zeppelins were slow and sluggish machines, vulnerable to attack. It was not long before pilots of the RFC flying from airfields in the Network began to bring home their attacks. As the losses mounted the cost of replacing crews and equipment was considered too high and gradually the threat of Zeppelin attack subsided. After that first raid on London, killing 13 and injuring 87, fifty one raids were launched from four bases on the German mainland which in total exacted more than one thousand casualties.

Clearly aircraft were assuming an important role in warfare which, as the new century progressed, was becoming more and more mechanised. There was increasing use of aircraft at home and abroad and with it came the need to perfect tactics and master developments in aeronautical engineering. Recognising the need to train its fliers the Royal Naval Air Service set about creating a centralised college of flying. Constructed at Cranwell it was one of the largest airfields of the time, swallowing up three thousand acres of prime Lincolnshire farmland to provide a flying base where air and ground support crews could be trained to the appropriate standard in the procedures and techniques of their service.

Then, nearly eight months before Germany's surrender and the Armistice signing to end the war, the Royal Flying Corps and Royal Naval Air Service were amalgamated to form the Royal Air Force. The training establishment at Cranwell was inherited by the newly created service and used as its training base. By this time the RAF had a large force of up-to-date aircraft whose crews needed induction courses and briefings to fly to full capability.

Peacetime was for making aviation history and breaking records. It was a time when pioneers set out to test planes to their furthest limits. In 1920 the world air speed record stood at 194mph but nine years later it would be pushed to 357mph. A year later, in 1930, Amy Johnson would fly solo from England to Australia.

Flying became a sport and was popularised by Flying Circuses which toured the length and breadth of England. Often headed by fliers who had become military aces in the dark days of war, they stopped at even the smallest of towns to give exhibitions of flying and offer the chance of pleasure flights. There was never any need for full airfield facilities: just a flat area capable of receiving a biplane or two and a track giving access to the paying public and a ground support vehicle, most often a light pick-up carrying a few basic tools and cans of aviation spirit.

The visits of two Air Pageants to West Norfolk are remembered by Freda Allen who bravely took the opportunity of a ten shilling (50p) flight in what she later described as a 'spartan plane'. About a dozen planes arrived at King's Lynn flying cross country from an earlier stop at Thetford. They proceeded to provide displays of flying, parachuting and offer pleasure flights from a field known as Sayer's Marsh which had been especially cleared of ten trees and blackthorn bushes to allow flying to take place. It left the mayor of the time impressed enough to hope that the field could become a municipal flying ground.

Later, Freda would travel by car to Heacham where Sir Alan Cobham's Great Air Display was scheduled to provide "Afternoon and evening displays continuous from 2.30pm. Admission 1/3d (about 6p), child 6d (2½p). Cars 1/– (5p), cycles 2d (about 1p)." Flights were available at the cost of 4/–, a bargain on the price charged by the earlier Pageant which was raising money for the British Hospital in London. During the display she remembers Flt Lt Geoffrey Tyson carrying out the stunt of picking up a handkerchief with the wingtip of his aircraft while in flight. Then, inspired by her experiences there and at Sayer's Marsh, she paid for a pleasure flight which, completely unknown to her, included a number of aerobatic rolls and manœuvers. "Friends who were my car passengers said that I looked green! And I drove them home to Lynn – the poor things must have feared for their very lives even though traffic was sparse in those days," she recalls. Two days later Sir Alan Cobham's machines were flying over Hunstanton and out over the Wash en-route for Wakefield, Yorkshire.

Aero-clubs began to open throughout the country catering for leisure flying. Municipal airfields developed to provide services to local industry and commerce. Flying became a spectator sport and large crowds were assured whenever Cobham or any other touring group gave an exhibition. In 1934 sixty four planes set off in a race from Mildenhall airfield in Suffolk to fly to Melbourne, Australia, where the winners would be met by Amy Johnson.

In that same year the Royal Air Force announced an expansion of its operation to meet the threat being posed by Germany which was fast rearming its own forces. Adolf Hitler made himself a dictator in 1933 and believing that war wearied western powers would not use any force to stop him, began to re-equip and rebuild Germany's military power. In 1936 he defied the Treaty of Versaille by marching his troops into the Rhineland which Germany had agreed to leave demilitarised. In 1938 he seized Austria and on September 1, 1939, invaded Poland.

* * * *

On that fateful day a pilot, air gunner and a naval officer acting as observer, were placed on standby at RAF Wyton. This airfield bordering the shire town of Huntingdon had grown from a small RFC training site into a massive aerodrome during the 1934 expansion programme and now, five years on, it seemed the work of enlarging and re-equiping would soon be justified.

At the precise moment the Prime Minister began his radio broadcast to the waiting country and the world, a War Telegram was despatched to a number of military chiefs authorised to receive it. It reached the Commander of RAF Wyton relayed through Group Headquarters and while he read it local people in houses around the airfield and in the nearby town tuned in to radios to hear Neville Chamberlain tell them that an ultimatum had been given to Germany to stop aggressive action in Poland. If satisfactory assurance had not been received by 11am that day a state of war would exist from that time. At 11.15 the country heard the news that 'no such undertaking has been received and this country is at war with Germany.'

On the airfield at Wyton the crew who had been standing by for three days prepared to take off. Half an hour after receipt of the War Telegram at the station, a Blenheim bomber with its crew of three was airborne on a mission to photograph units of the German naval fleet. Flying high in extreme weather conditions the plane's wireless set could not be operated and contact was lost until it returned to base at 4.50pm that afternoon. That evening an entry was made in the squadron log book: "Duty successful. 75 photos taken of German fleet. The first RAF aircraft to cross the German frontier."

Britain was at war with Germany and the Royal Air Force was operational within the first hour. Next day, missions would be flown to bomb the German navy. Air war was under way with a vengeance and in order to sustain the effort, further expansion of the Royal Air Force would be required. Not least the number of flying fields would need to be increased. In the traditional heartland of early flying, the East Anglian counties of Norfolk, Suffolk, Cambridgeshire and parts of Huntingdon and Bedfordshire, there were only fifteen active military airfields when war broke out. By May 1945 there would be 107.

* * * *

The standard design adopted for bomber stations throughout the war was the Class A type which consisted of three intersecting runways. The main runway was 2,000 yards long, with two subsidiaries of 1,400 yards each. The main runway was aligned with the prevailing wind and about fifty feet wide. Joining the ends of all three runways around the outside would be a perimeter track. Leading from this, outside the circle created by the perimeter track, were hard standing dispersal points where aircraft would stand. With a circle of concrete laid to permit turning and a narrow access strip from the perimeter track they were arranged singly or in pairs and affectionately known as 'pan handle' or 'spectacle' dispersals according to the shapes they made when viewed from the air.

Fighter stations needed shorter runways and often they would consist merely of grass strips, or grass overlaid with metal tracking. Hard standings were provided so that repairs and maintenance could be carried out and many had a concreted perimeter track circling the grass or tracking runways of 1,300 yards for the main, and 1,100 for the two intersecting subsidiaries.

The stations would consist of buildings which will be considered in greater detail later. Broadly they were arranged around two main sites. The technical site provided home to the familiar shapes of control tower, hangars and workshops and was always close to the runways. On the domestic site, often a distance away among trees and copses surrounding the area cleared for operations, would be arranged living quarters and places for leisure and recreation.

Near to the main stations would be satellite bases. Early on these were little more than fields with runways. Any accommodation for men and machines was in tents or temporary structures. Satellite stations would be used for training crews from the 'parent' field where their activities would not interrupt more urgent tasks. At critical times they were used as take-off and landing sites during raids. But as the need to expand the service quickly and effectively became apparent, it was realised that satellites could be upgraded. With accommodation for air and ground crews and concrete runways they could become homes to squadrons in their own right. Rapidly they ceased to be secondary extensions of an airfield nearby to become primary combat units.

But curiously enough, not all airfields constructed at that time saw active flying. One station in Cambridgeshire only came alive at night when lights came on. No flying ever took place and there were few service personnel ever seen near it. Small wonder that local people were confidently predicting that it was a classified site where secret weapons were being developed which could only be tested during the hours of darkness.

The truth was nowhere near so fanciful. In reality it was one of a number of decoy sites where mock flare paths, mock aircraft and mock hangars had been arranged to confuse the enemy. On decoys which operated during daylight hours (known as K sites) aircraft and buildings were constructed of plywood and placed around a field which, from the air, gave the impression of an operational station. The British film industry at the time, with limited budgets to spend on film-making, turned its attentions to the war effort using its scenery workshops and illusory skills to construct 400 imitation fighters and bombers to add realism to the decoys.

Other sites operated at night (Q sites). Headlamps used to represent aircraft landing lights were moved about a field and in darkness there was little need for lavish props to add to the illusion. Eventually the lights were mounted on carriages guided by wires and propelled by cordite cartridges.

The idea of the K and Q sites was to draw the attention of enemy raiders away from nearby operational stations and to that end they were successful. By the end of 1941 the 190 decoy sites constructed had been the target of 350 enemy attacks. However, while Q sites could operate with a minimum of two servicemen, K sites needed at least two dozen to create the right impression of reality. It was an expensive use of manpower, particularly since the Luftwaffe began to be more discriminating about its targets and with the capture of a top secret map of decoy sites, soon knew which airfields were genuine and which were not. In 1942 the use of decoys was abandoned.

Along the eastern seaboard three emergency landing grounds were constructed to provide a place for stragglers or damaged planes who could not make it safely back to their home airfields. Situated at Carnaby in Yorkshire, Woodbridge in Suffolk and Manston in Kent they were places for landing and take off only.

At Woodbridge only one runway was constructed but this was 3,000 yards long and five times the normal width, divided into three lanes. The southern most lane was for emergencies where any plane could land without the need to contact flying control to obtain permission. At Carnaby a Fog Intensive Dispersal Operation (FIDO) was built where petrol was forced along pressurised pipes beside the runway. At intervals along the pipe, burning took place and the heat from the burning fuel and the light it created served to disperse fog and mist and mark out the runway at the same time. The three emergency landing grounds were staffed with only minimal personnel whose trades and skills were in emergency medical aid, fire-fighting, radio communication, salvage and recovery. Facilities were primitive on these airfields but conditions hardly mattered to the thousands of crews who made landings there after limping home across the channel.

A few airfields were used as storage sites for aircraft. At RAF Kemble in Gloucestershire 500 planes were stored prior to allocation to squadrons and airfields around the country. Even on a non-combat station, life was not always easy. Many of the planes at Kemble were stored camouflaged under trees where they were prone to damage by winds and frost. The ever present danger of sabotage had to be countered and large numbers of men made up guard parties to ensure that the entire stock of planes was being watched.

So, as the pattern of war unfolded and the Royal Air Force grew in size and power, airfields, satellites, decoys and emergency landing grounds began to appear, almost overnight, in hitherto peaceful countryside. In fact, in Norfolk the concentration of airfields and satellites was so great that in the north of the county it was impossible to drive for more than ten minutes without skirting an aerodrome.

Above *Zeppelin raids on mainland Britain led to the establishment of airfields in a defensive line across eastern England in the Home Defence Network. This picture shows damage to a house in King's Lynn only a few miles away from one of the first military airfields at Narborough.* (Author's collection)

Right *Mess facilities on an early satellite station where meals were transported from the 'parent' airfield and eaten in the open.* (Bill Muddimer)

Below *Caught with his pants down!*
The speed to complete an airfield was often such that servicemen moved in before contractors moved out. Here an airman uses wash-house facilities comprising tent, tank and towels! (Bill Muddimer)

2 Construction

The first indication of an airfield being built in the vicinity was quiet, belying the noise and activity which would ensue as construction got under way. Through letter boxes of land-owners and tenant farmers, buff coloured envelopes arrived bringing letters from the Air Ministry. Citing passages from the Defence of the Realm Act, the letter proposed the taking of land for construction of a military airfield and assured the recipient that appropriate financial recompense would be forthcoming.

At a farm on the northern edge of Suffolk where lush farmland borders the bleak forested wastes of Breckland, the requisitioning letter arrived in the mid 1930s when the Ministry began construction of an airfield as part of the RAF expansion programme. Land from the parishes of Fakenham Magna, Troston, Honington and Rymer would be taken to create a village larger than anything nearby, to be known as RAF Honington.

John Wortley was a tenant farmer when the letters arrived. He spoke favourably of the settlement price reached: 'The Air Ministry were very fair in their dealings, as they should have been considering that about a third of my livelihood was gone.'

Once terms had been agreed parties of surveyors arrived in small, black saloon cars and navigated their way along lanes and cart tracks into the deepest corners of the local countryside. There they set up theodolites, took readings and made borings for soil samples. Diggers and graders followed to clear the newly acquired land of trees and hedges, leaving local villages with a rich store of fuel for drying and burning. As was to become common on East Anglian sites, former marl pits had to be filled in before a level surface suitable for flying operations could be created. At Honington a miner's tramway was laid so that tubs of soil could be easily moved to where pits needed filling.

One day as labourers worked to change the landscape and carve on it a new chapter in its history, they found themselves disturbing the past deep underground. As they worked the site of the parish gibbett was unearthed and there, still intact inside, was the skeleton of its last victim. He had been a local man, sentenced by magistrates to hang until dead for the murder of his daughter. Among the first on the scene of the newly discovered relics was John Wortley who recalls that the victim's boots were still intact. So too were his teeth in the jawbone, although after a short period of exposure to the chill twentieth century air they dropped out, falling on the mud around.

From then on rapid progress was made and local farmers watched the airfield rise as they worked their land. Soon substantial hip-roofed hangars had gone up constructed of brick, steel and glass, and marking the horizon with their bulk. No farm building at that time matched their sheer size and height. But even these structures would contrast with the corrugated steel buildings erected when the sense of urgency was heightened and conflict was near or already underway.

Runways at Honington were first laid by a curious process. Bags of cement were laid flat on the ground and were then driven over by a spiked roller. This punctured the bags and allowed their contents to mix with soil and rainwater to create a surface strong enough for planes in service at that time.

The airfield increased its size on two further occasions and when the next major land purchase occurred in 1942, the constraints of a wartime economy had made shrewder negotiators of the Air Ministry men. Settlements made were less generous than those in peacetime.

When war was underway and Honington was in use as an operational station there were restrictions placed on activities around it.

In fields which bordered the ends of the runways John Wortley was forbidden to build a hay or straw stack higher than twelve feet. He was required to report the use of all tall machinery, such as threshing tackle and elevator, to the station's flying control before work could commence.

* * * *

Those airfields built during hostilities had a greater urgency about them and the impact on the quiet rural countryside was more obvious. The need for the aerodromes to be completed according to strict schedules filled narrow lanes with heavy lorries carrying construction materials. Backroads which had been used to the sedate pace of horse drawn carts and the occasional Fordson tractor suddenly echoed to lumbering Foden steam wagons bringing cargoes of cement for runway construction. Yet before runways could be laid – and more advanced techniques were now required than those used at Honington a few years earlier – firm foundations had to be made. Many a local quarry had its annual turnover boosted as truck after truck queued for aggregates to take back to the construction site.

One driver at the time was Colin Acock who worked for the firm of John Laing. Many of the main contractors were household names in building and civil engineering: Richard Costain, Bovis, Taylor Woodrow and Humphrey's among them. It was a busy time for the drivers who worked for the successful tenderers.

"All routes were planned and strictly adhered to as many of the airfields were situated amid agricultural land and country lanes were not suitable for two lorries to pass each other," recalls Colin Acock.

Occasionally a driver would be keen to carry extra loads and earn a bonus in his weekly wage packet. Such a driver would often decide that, in his haste to carry more loads, the specially imposed one way systems did not apply and would often double back the way they had come. When this happened frequent jams occurred while drivers on the legitimate route argued with the renegade as to whose should be the vehicle to mount the verge – often with disastrous consequences, causing many a heavy truck to become stuck – or risk 'grounding' and overbalancing from taking to the bank trying to pass in the limited space available.

Given the deadlines and pressures on drivers, the nature of the loads being transported and the notorious twisting routes many had to take, it is perhaps surprising that there were so few accidents reported. Colin remembers two incidents. Returning to a construction site with six tons of wet ballast on board, one driver took a bend too fast. The lorry caught the grass verge and careered out of control into a tree where the impact forced the driver's cab backwards against the load, trapping its occupant and leaving him with severe back injuries. On another occasion, demonstrating a pride in the vehicle characteristic of drivers of the time, one man was greasing the hydraulics of his tipper truck when suddenly, without warning, the raised body of the vehicle fell upon him, severing his head from his body.

Women found employment as drivers and were rarely if ever short of offers of help when a puncture or breakdown occurred. One lady driver remembered by Colin had a profitable sideline and whenever he came across a certain lorry parked at the roadside with another drawn up in close proximity and their drivers nowhere to be seen, he always felt reasonably certain they were not bird nesting or gathering primroses!

* * * *

Back at the construction site teams of labourers would be at work laying bricks for billets, digging trenches to carry cables for landing light systems or levelling hardcore prior to the laying of concrete for perimeter track and runways. For this latter purpose vast amounts of concrete would be needed and a batching plant would be set up on site, with the raw materials brought to a steam driven, over-sized, concrete mixer capable of producing several yards of material with each mixing.

Building sites were cosmopolitan places with ethnic groups attracted by the prospect of a short spell of intensive well paid work. A gang of twelve carpenters from Greece, only two of whom spoke English, made regular appearances on a number of Norfolk airfield sites during the early years of war. Irish, Scots and Geordies all converged on the foreman's hut but even then there were often insufficient numbers of workers to get an airfield built on time, or keep concrete moving from the batching plant before it had a chance to set.

To increase the labour force a bus would be driven to the nearest large town or city and people would be recruited direct from the local labour exchange on a casual basis. Many coach operators whose fortunes had slumped when petrol rationing caused curtailment of excursion programmes, suddenly found themselves in demand as their smart, comfortable coaches were pressed into service as construction crew buses.

The system of using casual labour on construction sites was open to abuse and it was not long before local spivs were realising that money could be made for very small outlay and virtually no effort. On arrival at the site, casual workers would visit a stores shed or site hut and turn over a small token hooked to a board. Each token had the name and number of a particular worker and during the day the board would be checked to verify who was on site, so that pay packets could be prepared. On arrival at the site they would 'sign on' in the shed and turn their tokens on the board before making their way stealthily back to the waiting bus, whose driver had been tipped not to return too soon. Once on board the bus, the unscrupulous workers would hide behind seats or lay on the floor away from the gaze of any foreman, and would return to the city or town for the remainder of the day. When the same coach returned at the end of the day's labours, it would have the workers secreted on board, travelling back to the airfield to turn their tokens back and claim payment. Thus a day's wages would be amassed while the day remained clear for lucrative dealings on the local black market. The only cost of this deception would be a few hours travel on a coach and a financial consideration to the driver and any other workers on the site who would provide an alibi if called upon. It was a lucrative system but one which contractors eventually discovered and tried to clamp down on.

When he left school Peter Cron joined the construction firm of Richard Costain and worked on a number of airfield sites as a stores boy. He retains vivid memories of the labourers he came across, including one local farmer who turned away from husbandry of the land in favour of building upon it.

"Farming was practically at a standstill at this time. It was just before the Dig for Victory campaign. He was always known as 'Farmer' and when he was asked to use a pick and shovel by a foreman he asked if he would care for half a pound of butter. From then on Farmer became number one in charge of a concrete gang laying the perimeter track," he remembers.

"I had to go and check the disc tokens on the board and see if it tallied with the number at work in the gang. Farmer would tell me 'All present and correct except for one gone to the loo, one gone to the canteen for fags and one gone to look for both of them.' It was my job to find out if the men who were away existed on site or just on paper.

"When I arrived where his gang was working their small concrete mixer wasn't going. Farmer and the gang were sitting smoking happily, waiting for the plumber and his mate to bring another length of pipe to keep them supplied with water. I had a feeling it was another gentlemen's agreement – or perhaps half a dozen eggs."

At another site worked on by the stores boy there was a foreman whose heavy oilskin mackintosh and bowler hat, together with the book and pencil he always carried, created a shape on the horizon like a huge crow as he stalked through the mud, checking on progress. So distinct was his outline in the distance that watches would be posted and early warnings given so that any illicit games of Crown and Anchor would be cleared away long before he reached the gang.

But it was not all idle inactivity or deception, drinking or cards. The over-riding impression Peter Cron retains was of the speed at which construction proceeded and the rate at which the airfield took shape. During quiet periods in the stores hut he would watch the level of bricks between windows in barrack blocks visibly rise skywards and within seven to eight months all the buildings necessary for a fully functional airfield would be complete.

* * * *

The invasion of peaceful countryside by airmen and aircrew was presaged by that of building workers from other counties, and in some cases, other countries. In Norfolk the larger than life figure of Geordie Robertson is still remembered for his visits to the local post office each week to buy two sets of money orders for the families of the two wives he was reputed to have back in Tyneside.

There was an Irish gang who regularly commandeered a forward dump truck to drive between the site and nearest public house. The noise of the engine chugging and disturbing the peace at lunchtimes and summer evenings lead to complaints being made to the local constable. He apprehended the dumper's progress one day and warned them that it was a grave offence to misappropriate government issue fuel to drive an empty dumper. The Irish left aboard their trusty workhorse, crestfallen, without visiting the local hostelry. That evening the unmistakable noise of the dumper was heard again and when the policeman went to investigate discovered the vehicle laden as usual with Irish. But in the dumper section were three shovelfulls of sand. "Now catch us for driving an empty dumper," challenged the driver defiantly.

But the regular excursions to the public house were curtailed in a way neither villager or law enforcer would have wanted. Not long after, the dumper was making its regular trip with five or six Irishmen sitting and hanging from it. Negotiating a curve too quickly one of the passengers lost his grip, over balanced and fell off, meeting his death beneath the vehicle's wheels. Never again would the drinking party's dumper disturb the village peace.

In the early days of war when airfield building was beginning, Ireland was full of agents acting on behalf of English companies, charged with recruiting labour. One such recruit from County Cork, was Diarmud Collins who at that time was working in his uncle's business testing and packing eggs, and driving to collect them from outlying farms. The offer of a guaranteed wage of 2/– (10 pence) an hour was attractive to the thirty three year old Irishman, particularly since he had been assured that labour recruited locally on the United Kingdom mainland would only be receiving 1/5¼d (about 7p) per hour. The contract was signed and on August 22 1942 he arrived at Liverpool for Diss in South Norfolk. The train journey was broken overnight in London where Diarmud and his 38 countrymen stayed at University College.

At Diss railway station they were loaded into a lorry and taken to a site four miles north east of the small market town where work on constructing an airfield at Thorpe Abbotts was already well underway. Footings for runways were already dug and temporary barracks had been built to house the influx of workers which, at one point, numbered 2300 people. He quickly rose to the rank of foreman and supervised his group of fellow countrymen laying bases for Nissen huts, making paths or laying drains and sewers.

The Americans who were to fly from Thorpe Abbotts arrived soon after the first runway was constructed. Flying operations concentrated on Number One runway (the main) while construction work continued on the two subsidiaries. "When they arrived it was as if all the wolves of America had been let loose," Diarmud recalls. "They didn't know where they were. Nobody had bothered to tell them at any stage on their journey. They climbed off the lorries and walked around the place howling, shouting and making as much noise as they could. They just could not believe how peaceful and quiet it all was."

Construction work took second place to flying operations and many were the occasions when the Irish labourers working on reinforcing or resurfacing runways and taxiways had their work postponed while aircraft took off and landed on missions or training flights. There was such a backlog that labourers worked at night to unload lorries of materials required for work the next day. These occasions were rewarded by handsome bonus and overtime payments which helped sweeten the expectation that they would all be reporting for duty later that day, just as if they had spent the night asleep in their bunks.

At that time it was work, work and more work. Diarmud Collins' main recreation at the time was a Saturday afternoon trip to Diss. There a shopping expedition would take place at the local International Stores and the post office, where postal orders would be purchased and sent to the family back in Ireland. After that there was a visit to a local cafe and then a bus trip or walk back to the airfield. Some weekends he would travel to London to meet his childhood sweetheart and fiancée who was working as a nurse in Kent. They would spend their weekend at Hyde Park, Kew Gardens or in the Irish Club where the resident band played through many an air raid.

After his first six months he returned to Ireland for a holiday, boasting to his family that back in England he was the owner of two motorcycles. He rejected offers to stay in his homeland and returned to Thorpe Abbotts and continued playing his part in laying two and a half miles of surface drains, twelve miles of surplus water drains, 35,000 yards of tarmac, four and a third million bricks, eleven miles of roadway, five miles of water main and three and a half miles of sewer. When all the costs had been accounted, John Laing drew up their bill: 'Full wartime aerodrome with runways and perimeter track but buildings of temporary construction: £784,589.0s.0d.'

* * * *

In the early days of the war some runways were made of metal tracking laid on grass and held in place by large steel pins. The tracking allowed the grass to grow through, giving all the advantages of a grass air strip but without the attendant risks of the muddy quagmire during winter months rendering many grass airfields inoperative. At Great Sampford in Essex a gang of four hundred men laid 1,000 yards of metal tracking in just 12 hours to provide a runway for fighter aircraft.

Often airfields would be built with the emphasis on runways or hard standings and at some stations aircraft were housed in hangars while their pilots and ground crews lived under canvas. At Bodney in Norfolk station personnel had wooden huts to live in but still had to draw their water from a nearby stream until a water tower was built on higher ground and the base provided with its own piped supply.

The development and expansion of airfields did not remain static after building. Even when a station had been handed over for operations it was possible builders would be moving back at some time during its life to rebuild or enlarge it. In some areas construction techniques had not been properly supervised and evidence of shoddy workmanship became apparent, which needed to be made good. All the time aircraft manufacturers were engaged in building larger and larger aircraft, capable of carrying even heavier bomb loads. Such developments had implications for the runways which would need strengthening or extending. Airfields also changed status from fighter to bomber stations, often creating a need for hard runways to be constructed to replace grass landing strips or metal tracking.

It was a time of frantic activity and at its height in 1942 a labour force of more than 60,000 men was engaged solely in the building of airfields. During the period from September 1938 to September 1946 a total of 560 airfields in the British Isles had been constructed, extended or upgraded.

3 The Buildings

When construction work was nearing completion the site would begin to be occupied by service personnel. What they inherited as the labourers moved out varied from station to station. In some cases it was virtually a complete town of domestic and technical buildings, some camouflaged against the surrounding countryside, with runways and a network of roads and taxiways. At others the set up was much less plush and, like Bodney in Norfolk, still reminiscent of the pioneering days. It consisted of a simple field with no hard runways. Wooden huts were sited either side of the landing field, A flight on one side and B flight on the other. An ammunition dump was sited at the foot of the landing strip near a main road and Blenheim bombers stationed there were drawn up on each side of the grass landing strip when not on operations. Until living accommodation on the site was extended, ground crew were ferried in each day by road from nearby Watton.

Bodney was a war built site and contrasted with Bircham Newton in the same county. Bircham had its roots in the earliest days of the Royal Flying Corps and by 1932 had already been earmarked a heavy bomber station. During the 1930s expansion period considerable work took place at the airfield, with large hip roofed hangars constructed to replace the earlier wooden sheds. Living quarters built in the 1920s were in mock Georgian style, each block named after a key battle in World War One. By the time of expansion in the mid 1930s austerity was the key and spartan brick, flatroofed barracks appeared beside the warm looking, gable ended constructions of a decade earlier.

The day to day needs of airfield personnel were met in metal Nissen huts, wood built hutments or brick buildings, the latter often coated with an exterior finish of concrete known as Maycrete. These structures accommodated canteens, squash courts, cinemas, plus the usual offices and barracks. Around the domestic site were air raid shelters. All that would be visible above ground was a grass mound with a cutting slicing it in two. This, strengthened on each side by concrete walls, provided access to an underground shelter where several more inches of concrete, metal and earth afforded a measure of protection from anything save a direct hit. The same subterranean construction epitomised the bomb store – although this was on a larger scale than air raid shelters. Because of the ever present danger of explosion the bomb 'dump' was usually sited on a side of the airfield furthest from where people were living and working.

Chief among structures on the technical site were hangars. Those built before 1939 were imposing structures, providing bays to accommodate up to ten large bomber aircraft. Before hostilities they were in use as 'home' and workshop – with aircraft being serviced and repaired there, and stored when not in operational use. As enemy raids on airfields increased it was realised that one well placed bomb on a hangar could destroy much of a squadron's strength. Aircraft were then stood at dispersals around the perimeter, often concealed by trees, and only went to hangars for repairs. Smaller, lighter hangars appeared on war-built airfields constructed predominantly from steel sheets and featuring curved or hipped roofs according to type and manufacturer.

Within easy reach of the hangars on some stations were large open ended garage-like buildings where aircraft engines could be placed on test rigs and slung for 'running tests'. Not far away would be the control tower. Originally small wooden sheds, by the outbreak of war they had developed into two storied buildings with flat roofs and a balcony on three sides. Many had the shed like structure, or watch office, perched atop the flat roof, which afforded a view of landings and take offs over much of the flying field. Here were based the predecessors of the modern day flight controllers, station met officers and radio communications staff. Close by was accommodation for the station fire crew – a central location giving fire and crash tenders access to incidents anywhere on the airfield. Ambulance stations, sick quarters and mortuary were usually sited as far away from the operational area as possible.

A few airfields possessed a concrete dome construction, looking for all the world like some gigantic pimple. They were rare enough for opinions to differ regarding their use. One theory is navigational training took place inside them, with star groups projected upon the inside dome as in a planetarium. A different use is outlined by Ian Massingham who, as a member of the Air Training Corps from 1942–44, had the chance to visit a number of airfields including one at Langham, Norfolk, which still retains the dome trainer structure.

He remembers it being used to train ground gunners. Projection equipment inside the dome displayed films of dive bombers making attacks from various positions from the 360 degree interior of the dome. The image was moved around the dome by a complicated system of cams and mirrors. Films were projected in colour with full sound creating a realistic experience for the trainees sitting in the middle of the dome at an anti-aircraft gun.

It is not always appreciated that the sudden arrival of an airfield in a rural area had implications for water and electricity supplies. An airfield covering between 700 and 1000 acres with a thousand or more people living and working on it was frequently larger than local villages. Many communities still did not have piped water or access to power supplies. So, in order to prevent excessive demands being made on local services where they existed, and in order to provide public utilities on airfields where they did not, water towers and generator stations had to be built.

On wartime stations water towers were simply large galvanised tanks supported in the air by a structure of metal supports and legs. During the expansion period the tanks were enclosed inside brickwork and designs for them had been approved by the Royal Fine Arts Commission. In either case, a pump house would be nearby to circulate water around the domestic and technical sites. On some expansion airfields there was even a boiler house and central heating station to circulate hot water through the pipes and radiators of a station's barracks and offices.

Station headquarters was housed in offices near the main airfield entrance, which itself was usually dominated by some form of guard house. This provided a base for checks on personnel and vehicles entering or leaving the site, and

Left *With camouflage helping it blend in with the rural landscape below, a Halifax of 192 sqd. sets out on another operation.* (A.J. Berry)

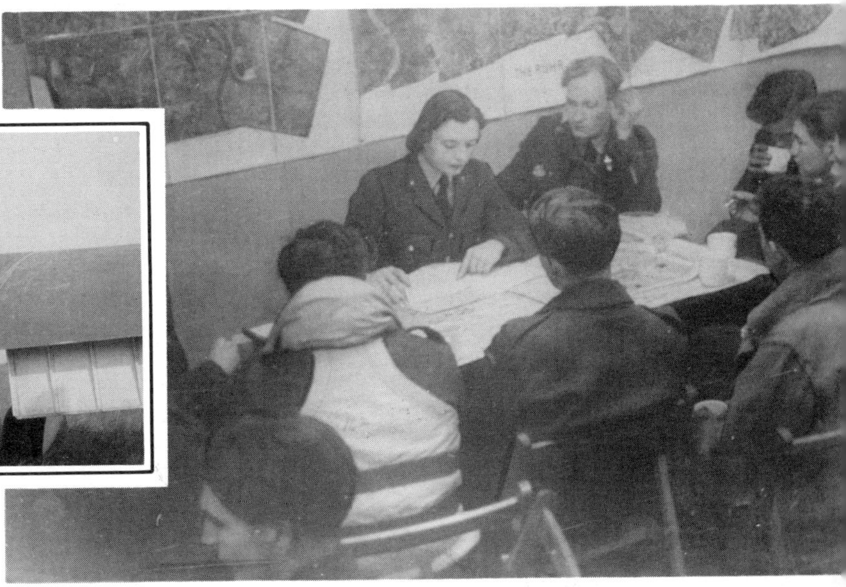

Above *Canon fire from an enemy aircraft damaged the wing and flap assembly of this Hurricane which still managed to limp back to its home airfield.* (Edith Kup)

Above right *Returning crews report their mission to Edith Kup during a debriefing. On the wall behind them are photographic mosaics of Hamburg and the Ruhr district.* (Edith Kup)

Right *While the squadrons are aloft senior staff and WAAF members collate reports in the plotting room.* (Edith Kup)

provided cell accommodation for any erring servicemen to cool off in. Somewhere on site, well away from other buildings, would be a Radio Direction Finding station. Its single storey flat roofed building was protected by a barbed wire fence but the radio mast above always succeeded in pinpointing its exact whereabouts.

Thus the main buildings comprising an airfield can be accounted for. Add to them a place for the Photographic Section to process and interpret the results of reconnaisance flights before and after raids; build a store for packing parachutes, a tailor's shop and food stores; construct compounds for airfield vehicles, fuel storage areas and a decontamination block in the event of gas attacks and it becomes apparent how advanced airfields had become since the pioneering RFC days just 24 years earlier.

When construction workers finally vacated the site, they left behind not simply an operational air station. Whether the accommodation was in the grand manner of pre-war days, or the hasty and utilitarian constructions of wartime, it was the home for a community to live and work in. Offices, workshops, mess halls, billets, hangars and stores, all were created with the necessary infrastructure to deliver power at the flick of a switch, water at the turn of a tap and, of course, an empty toilet pan at the yank of a chain. Many airfields represented the most ambitious civil engineering and building projects ever to come to many rural areas. Villages which had evolved slowly through history and still had some catching up to do, found themselves neighbours to new communities which arrived in a matter of months, built from scratch and which possessed all modern conveniences.

4 Ops

Mere buildings do not make an airfield, nor the personalities of those who work there. Bind them together in a common purpose, the operation of delivering blows to the heart of an aggressor, and an airfield comes alive. In this chapter a visit is made to an anonymous airfield 'somewhere in England'. The year is 1942, May has nearly run its course and bad weather is beginning to show signs of improvement.

Paths between the Nissen huts are puddled and bicycle wheels have left deep muddy ruts in the grass around the billets. There are bikes everywhere: they are the means of mass and rapid transit around the 'drome. Each squadron has its allocation but it hardly seems to matter. When there is a need to travel quickly to the mess the nearest roadworthy machine is pressed into action regardless of ownership.

From a Tannoy attached to a telegraph post a tinny, staccato voice reminds that flight briefing will commence at 1600 hours. The door of a Nissen hut swings open and two men in air force blue run out, hurrying off in the direction of the briefing room. They had a late lunch in the mess and do not welcome this intrusion into their idle afternoon, but at the same time are glad the briefing has been called. They have been kept waiting in a state of readiness for too long. By now they are keen to finish the job. The weather is clearing over the airfield and the suggestion is that it may also be clearing over the target. It looks like operations tonight.

In the briefing room they sit around tables, the better to make notes. Before them is a dais, a blackboard and ahead of them on the wall a large length of blackout curtain is obscuring something. On the opposite wall a poster proclaims: 'It is better to keep your mouth shut and let people think you are a fool than to open it and remove all doubt.' A squeaking door at the back heralds the arrival of the station commander and other senior staff. Silence descends upon the gathering, broken only by the sound of boots in rapid stride echoing hollow on the hutment's floorboards.

The Station Commander pulls at the blackout curtain and reveals behind it a map of Europe. He announces that their target will be Cologne, the very heart of the German homelands. He tells his squadrons they will be part of a force of more than one thousand bombers and they will be delivering a decisive blow at the enemy's industry. He then hands over to the station navigation officer who provides large scale maps of the target and gives precise information and timings for their arrival on the Suffolk coast. There squadrons from 55 other similar airfields will form up into a dark cloud of menace before proceeding to enemy occupied Europe.

The station met. officer forecasts the weather over the target area and indicates conditions along the flight path. Crews will need this information to balance out delicate fuel equations: avoiding storms, headwinds or deviating too far off course will soon use up limited supplies of aviation spirit. Intelligence officers follow and share the knowledge they have of the enemy's night time defences, including the whereabouts of known night fighter stations which will probably be alerted to impede their progress.

On the blackboard the signals officer chalks up details of radio frequencies to be used and the armaments officer tells the size, type and number of bombs they will be carrying. There is time for questions before the commanding officer rises to his feet and steps forward to address his men. He reads to them from a letter from Sir Arthur Harris, Commander-in-Chief of Bomber Command.

"The force of which you are about to take part tonight is at least twice the size and more than four times the carrying capacity of the largest air force ever before concentrated on one objective," he reads. "You have the opportunity therefore to strike a blow at the enemy which will resound, not only throughout Germany, but throughout the world. In your hands be the means of destroying a major part of the resources by which the enemy's war effort is maintained. It depends, however, upon each individual crew whether full concentration is achieved. Press home your attack to your precise objectives with the utmost determination and resolution in the full knowledge that, if you individually succeed, the most shattering and devastating blow will have been delivered against the very vitals of the enemy. Let him have it, right on the chin."

The briefing is complete. Senior staff leave the room. Benches and tables scrape on the floorboards as crews head to tables arranged at the side where they will collect a chocolate ration for the night ahead, and if they are lucky, maybe a tin of pineapple juice to take aloft with them. From the briefing they walk to their squadron's crew room.

Flying kit is taken from lockers and crews change into it. Their flying uniform is similar in cut to army battle dress. A sweater is worn and an Irvine jacket which contains a 'Mae West' which will inflate instantly on contact with water. Silk socks are issued and over them stout woollen stockings are topped by heavy flying boots lined with lamb's wool. The flying helmet contains its own oxygen mask and a tube that will be connected to the aircraft's oxygen supply dangles from it like a small elephant trunk. The helmet also has its own built in microphone mouthpiece and earphones.

While in the crew room aircraft captains and navigators leave the rest, form huddles and conduct briefings of their own. They discuss the course they have been set and balance fuel usage against distance to be flown and bomb load being carried. The navigator takes a flat satchel where maps are stored. It also carries cartridges for the Very light signalling pistol and 'flimsies' made of rice paper on which are typed the procedure to be adopted if their plane is lost and requires wireless assistance. In the event of emergency, the instructions will be acted upon and the flimsies become extra crew rations before they can fall into enemy hands.

From the crew room they move across to the parachute section building. Not all aircrew will wear their 'chutes during the mission but they will never be far away, clipped to hooks near each work station on the aircraft. Thus kitted with clothing, rations, maps and target details they walk to part of the apron where a motley assortment of trucks and buses wait to take them to aircraft scattered at dispersal points all around the airfield perimeter, well away from the technical area where they are now.

* * * *

At dispersal huge aircraft await crews to bring them to life. Ground crews clamber over the airframes and make final checks. While the briefings went on there was much activity here in the open air. Bowsers visited each hard standing to pump fuel into the tanks of each 'plane, and tractors towing trains of bomb trolleys arrived and discharged their loads into the very bowels of the aircraft. It has also been busy in the hangars. Hard work there by technicians and mechanics has meant that a bomber which returned from its last mission with only one engine working, is now firing on all four again and can join the operation. The repaired plane has been towed from the hangar by tractor round the perimeter track to its dispersal and has been fuelled and bombed up along with the rest. Tonight the squadron will be flying at full strength.

Ground crews, idle after their checks have been completed, exchange banter with farm workers on the other side of the fence. It is peaceful, the ominous calm that always precedes the storm. Soon in the distance there is the characteristic whining of Bedford engines under labour: the crew trucks and buses are approaching.

While crew members clutching flasks of hot tea or coffee climb and swing themselves into the fuselage, the captains of each aircraft have a final meeting before becoming airborne. This time it is with members of the groundcrew. The list of problems which presented on the last mission is checked to see if all repairs have been completed. All work has been done, and one problem which the crew had not even reported has been diagnosed and repaired, notes a leading aircraftsman with pride.

The size of the plane standing at dispersal is illusory. Despite its apparent size from the outside, the accommodation inside is cramped and functional. The rear gunner crawls on hands and knees through the fuselage into his turret and closes the door between him and his colleagues. Small wonder that an air gunner once said: 'The striking thing about a tail turret is the sense of detachment it gives you.' The wireless operator goes to his work station near the navigator. They huddle in tiny alcoves with the impediments of their trade cramped on three sides around them. Up front the captain and flight engineer take the controls in the cockpit.

Within just a few feet of them all are tanks of high octane aviation fuel and high explosive bombs: a lethal combination which they will be responsible for piloting over land and sea. And while over enemy territory there will be the ever present threat of strafing by night fighters or running into flak from ground based anti-aircraft batteries. And on this mission there is also the risk of a mid air collision with another plane in the formation. They were told at briefing the possibility of such a collision was a thousand to one. Sitting in the aircraft they have a few brief moments to console themselves that there are 999 chances against it being them.

Once at their work positions checks are made. The wireless operator checks his equipment is functioning correctly and calls up the tower. Any communications he sends during the mission will be received there and it is vital that everything works and effective contact is made. On the flight deck pilots go through a list of mechanical and other equipment checks. Then when all is complete doors and hatches are battened and there is nothing left to do but sit and await instructions from the tower. It has been known to take hours before the order comes through and they are confined to the aircraft until it does. Today however, there is activity in the distance, near the technical site, which suggests that the order for take off will not be long in coming.

Near the control tower they can see the fire tender and station ambulance drive from their garages to positions of readiness. Personnel able to leave their duties are beginning to congregate around the hangar aprons where they can have a good view of the flying field. In ploughed fields around the perimeter work also stops: even away from the airfield there is a feeling that tonight's operation is something different, a little out of the ordinary.

The pilot is ordered to start and warm up the engines. They whine at first and blue smoke explodes from exhaust pipes and dissolves on the evening air. The propellers turn slowly at first, building up speed, windmilling occasionally until they are only decipherable as circles of blurred and spinning metal. Grass nearby flattens in the wind created. As the revs increase the whole aircraft shudders against the huge wood or metal chocks placed against the tyres to prevent forward movement.

Then they are pulled aside by groundcrew and the plane taxis slowly from its dispersal, along the pan-handle and on to the perimeter track which will lead it to the runway for take off. At every hardstanding this happens, huge giants edging forward slowly, soon to build up into a queue of roaring energy and fire power, each waiting for the signal to take off. It comes from a control van, easily distinguishable by its black and white checker board pattern, parked near the main runway.

A green light shows from the van and the leading aircraft has its throttles opened, moving forward, increasing speed all the time. Other aircraft move forward and take their turn. At regular intervals the green light shows and the process is repeated: between sixteen and twenty tons of men, machine, fuel and bombs hurtling down the runway and eating every foot of the 1000 yards needed to get them airborne.

The squadron is away. The noise of piston engines labouring to get their loads aloft mixes with cheers from ground crew as 'their plane' takes to the skies. The noise as they circle to make formation brings householders into their gardens keen to watch one small part of the air armada form up and begin its journey southeast. Children already in bed have their slumbers disturbed as the planes pass overhead with a heavy and menacing roar.

* * * *

Then there is silence. The silence of night and impending dark. The silence of tense anticipation, the quiet of waiting. There is little to do now on the ground but await the planes' return. Girlfriends offer up silent prayers for the safe delivery of boyfriends now many miles away; technicians and fitters hope that repairs they made were not completed too hastily.

In the control tower courses are marked on maps. A wireless operator maintains a constant vigil monitoring the alloted radio frequencies. Radio silence is always maintained while over the enemy coast but there might be an emergency broadcast or the need to check course.

By now an overall picture is beginning to emerge. A total strength of 1050 bombers has been sent over the target. Twenty seven planes have been stood down at the last minute with key crew members ill or with engines suffering from equipment failure. At one Yorkshire airfield two Wellington bombers have collided while still on the ground. In the home counties another Wellington failed to get enough height at take off and hit a tree at the end of the runway. Ninety more aircraft will abort the mission during the early stages before the target is reached and there may be some embarrassed looks on the faces of ground crews when they return.

Two hours and twenty minutes after taking off the first planes are discharging their bombs over Cologne. The time in Germany is half past midnight. The last flight to deliver its load will be over the target at three in the morning, just as the first planes are beginning to land back at their home bases.

There the results will be coming through and recorded on boards with coloured chalks or on charts by means of pins and labels. Senior staff step out from the watch office of the control tower and stand on the open balcony, the better to hear a distant roar of engines indicating returning 'planes. As soon as they are heard, instructions are broadcast giving permission to land, or the order in which to land if several are arriving together. Flare path lights are turned on for each landing and the station emergency services are back in position for their second vigil of the raid.

Not all landings are straightforward. Enemy early warning systems picked up the bomber stream as it formed, night fighters were scrambled and a number of aircraft were shot down. Heavy flak over the target area has damaged more than 80 aircraft and the returning force is vulnerable. It is no longer a concentrated mass wing of firepower, but is scattered and prone to attack from raiders. Some planes are obviously damaged. On others problems will not be obvious until they try and lower landing gear.

A message is radioed to the control tower. Emergency crews are put on alert and other circling planes are stacked and hold off while a damaged aircraft attempts landing. The pilot heads for grass beside the runway where the shock, impact and vibration waves of a belly landing can be reduced. Contact with ground is made. Propellors gouge out grass and earth before forward momentum buckles them and they cease turning. Without power the bomber skims over the ground, much lighter now than when it took off from this place. Slowed by the soft soil it reaches a resting place and slews round on itself so that the crew have a view of the furrow they have just ploughed across the airfield.

Heading for the crash site at speed are the fire tender and ambulance. Behind them follow trucks laden with men, anxious to reach the bomber's resting place and offer whatever help they can to rescue crew members. Others run to the crash site, some stand on the pedals of their bikes trying to overcome the resistance of damp, soft grass.

Apart from bruising and mission fatigue the crew are unharmed. They clamber out from an aircraft beached like a stranded whale. The bomb load has been discharged, no fuel pipes or electric systems are damaged and there is no possibility of fire. Personnel retire back to the hangars and tower area as the landings recommence. Later during the morning the recovery crew will bring its crane and expertise to bear, salvaging the 'plane and taking it into hangars for comprehensive checks and rebuilding.

Not all aircraft in distress have made it to their home bases. Messages reach the station from an East Anglian airfield who have unexpected guests: planes so badly damaged they could not complete their journey and had to land at the first opportunity. Some have been forced to ditch in the channel. Mayday calls have been received and the RAFs own rescue launches and local lifeboats have been alerted. All planes carried automatically inflating dinghies and crew members have a yellow covering to their flying helmets designed to help identify them from the air.

The planes which have landed without problem are met by trucks and buses that took the crews on the first stage of their journey to the enemy's heart. Now they wait to take them back to the briefing room where the mission began a day ago. The familiar furniture layout has been changed. Tables and chairs no longer face forwards but are set out around the room like a tea shop. In the corner on a trestle table are huge steaming urns of tea and coffee, piles of cups, saucers and plates, and stacks of sandwiches cut in triangles.

Crews are interrogated as they arrive. Over light refreshments, intelligence officers work through a questionnaire to build up a picture of the raid and the effectiveness of the enemy's air defences. And just as before while this room was busy, there is activity at dispersals too. Ground crews are already checking returned planes for damage and compiling lists of jobs to be done before they can be made fully operational again.

Then when the questionnaires have been completed the crews leave the briefing room and emerge into the dawning of a new day. In the mess breakfast is being served. They are offered good helpings of egg and bacon – food in short supply at this time and only available on limited ration to anyone other than aircrew. Tea in thick mugs precedes sleep. The move towards slumber is helped by the commanding officer who sends a signal to the mess that rum should be broken out in recognition of the squadron's successful part in the raid.

When their breakfasting is done and the last drop of bitter laced tea has been drained they will leave the mess and return to their billets. Outside full daylight is approaching. Around the airfield domestic site junior ranks are at work on litter patrol. The puddles are drying up and virtually all the squadron's cycles are themselves resting, double and treble parked, propped against hut walls. In the fields bordering the station land girls are arriving to begin their day's work.

5 Airborne

All airfield personnel living and working together created a feeling of community, sharing common experiences and working to achieve shared goals. But for members of aircrew there were other loyalties: they had their own closer community to be part of, that provided by their squadron. Then, in the air, each aircraft became a community in microcosm with every member fulfilling a key role in the small team that crewed it. The Air Ministry wrote lyrically of the life they shared when airborne in propaganda and morale boosting brochures about Bomber Command. "Their life is not that of other men – not that even of those in other branches of the service. For them nowadays much of the night is day, much of the day a time for sleep and repose. Discipline is constant yet flexible and its effect is to cause these men . . . to meet life as a familiar, sometimes an exacting, but always a common friend," wrote the Ministry men.

In the last chapter we watched as crews were briefed and planes took off for Cologne and, with the rest of an airfield community, we waited for their return. Let us now join a crew in the air and share with them some impression of nights that are days and see the work that takes place when the airfield is far, far away.

There is an eerie silence inside the aircraft; each member of the team has a job to do and there is no need for superfluous talk. Occasionally the steady drone of the four Rolls Royce Merlin engines is broken by the intercom crackling into life as the navigator pinpoints their position to the pilot. The navigator sits near the flight deck with maps and charts arrayed on the ledge in front of him. Sometimes he uses stars to work out the course, other times he uses wireless directional aids which can be used even when the plane is maintaining radio silence. He uses a torch as he makes calculations and plots the course. Behind him the wireless operator is making regular entries into a logbook. He works by the glow of an amber light hanging over the banks of receivers and dials that are part of the small cupboard, open at one side, where his flying hours are spent.

Loneliest of all the crew is the rear gunner, isolated in the tail turret. Cramped in a perspex bubble that protrudes beyond the tailplane, he sees nothing of the aircraft he is part of until the turret moves sideways. Always travelling with his back to the engines, seeing where the plane has been, never where it is going, he is linked to the plane's intercom and the sound of the crew exchanging brief messages. This is the only contact he has with the team of which he is an integral part.

They are now over enemy territory. Flying high to avoid anti-aircraft barrages the air becomes thinner and they don oxygen masks. As they near their target the ground batteries increase their fire power. In the darkness below there are momentary flickers of light as ground guns discharge their shells. Seconds later the air around them erupts as balls of fire and smoke burst on the darkness around them. Occasionally the shells are so close the plane is buffetted and shaken, but luckily not close enough for a hit. Concentrated beams of light rise from the ground now, criss-crossing the sky as searchlight crews try to fix on a target for the ground gunners. The closer they get to the target, the more intense the barrage becomes.

As the target nears the navigator goes forward to the bomb aiming position. When it is located he feeds back to the pilot that the time has come to make the bomb run. Below them the ground is ablaze with fires from incendiaries dropped by planes in the first attack wave. Bomb doors are opened hydraulically and from then the navigator is in charge of flying the plane, relaying instructions and directions to the pilot and co-pilot. The intercom is crackling even more and it is difficult to make out words. The RAF have allowed for this and when the instruction to turn right is given, the word 'right' is said only once. When the move is in the opposite direction the word is repeated: 'Left, left.' Even when the quality of inter-communication is poor the pilot can fly the aircraft by listening for single word or repeated directions.

The pilot holds his aircraft on course until the navigator reports "Bombs away". Navigator cum bomb aimer has not finished his work yet for he must watch the bombs fall and operate switches to control the rate at which they leave the aircraft – singly or in a stick of several. When the load has been discharged he reports "Bombs gone" and the pilot opens the throttle and operates the hydraulics to close the bomb doors. The aircraft turns for home but their task is not ended yet.

There is still flak to contend with and as they head towards home they are picked out in the beam of a searchlight. Within seconds other lights on the ground cluster home in on them. Shells explode close around but on this trip they are charmed and nothing touches them. Wireless silence is maintained and the air gunner keeps vigil watching the skies for any evidence of fighters preparing to attack.

One Wellington bomber in the force is attacked by night fighters over the German–Dutch border. The rear gun turret is shattered and its operator is wounded. The fuselage of the aircraft is on fire and the second pilot has been killed. As the plane limps toward England mechanical problems occur and without any switches being operated on the flight deck, the undercarriage and bomb doors drop open. Airspeed is reduced as the plane encounters resistance. The fire inside catches pipework and tubing, filling the work areas with the heavy acrid smell of burning rubber. Just as it seems nothing else can go wrong another fighter comes in to attack, forcing the bomber to fly at ground level for the remainder of the journey back. Overwhelming odds for a safe return are overcome and although unable to return to its home base in Lincolnshire the pilot makes a successful emergency landing at Honington.

Our crew are fortunate and reach home without incident. Waiting for them are crew buses that will take them to debriefing and interrogation. From the reports they provide and from similar intelligence collated from squadrons all over the eastern seaboard a picture emerges of RAF losses. Twenty nine bombers have been attacked by fighters but air gunners are claiming four shot down, one destroyed (subject to confirmation) and four others damaged. Of the bombers which have returned in a damaged condition only twelve are the result of fighter attack, the rest attributable to anti-aircraft fire.

 Soon after, another large scale raid takes place on Hamburg. As the Royal Air

Force heads towards Europe the Luftwaffe is flying from the fatherland to bomb Coventry in reprisal for Cologne. When our crews return to the briefing room from the Hamburg raid there is another message from Sir Arthur Harris: "By your skill, determination and courage in last night's operation you have undoubtedly struck the enemy a stunning blow. All and more that was expected of you, you have achieved." Prime Minister Winston Churchill sends his greetings to the crews: "I congratulate you and the whole of bomber command upon the remarkable feat of organisation which enabled you to despatch over a thousand bombers to the Cologne area in a single night and without confusion to concentrate their action over the target in so short a time as one hour and a half."

It is indeed an achievement which all personnel from the airfield community can be proud of.

* * * *

Not all the work of crews was geared to dropping bombs, nor was it always as spectacular or solicitous of praise as the Thousand Bomber raids. There were other tasks and early in the war aircraft dropped 'white bombs' – millions of leaflets and pamphlets released over enemy territory and occupied cities bringing propaganda messages from the British government. Such early raids were used by the RAF to supplement its intelligence gathering operations and served to familiarise crews with key areas which would subsequently become targets for bombs. Some fliers and observers argued that leaflet raids demanded an even greater level of skill and endurance than high explosive raids. Leaflet raids were carried out in all weathers and since they were lighter than bombs or incendiaries complex calculation and experimentation had to take place to deliver on the target areas – which were often small communities, restricted in size. Direction and strength of wind together with aircraft altitude became as important dropping paper from on high as it did when the load was high explosives.

Other tasks undertaken by squadrons included radar counter measures and mine laying. Long strips of metallised paper, like present day cooking foil, were dropped by the thousand to disrupt and confuse enemy radar. Minelaying called for accurate flying and navigation. Waters around Norway, Denmark, the Baltic, the Kiel Canal and Bay of Biscay were among those mined by bombers converted to carry parachute mines. These could only be released from low altitudes. Dropped from too great a height the parachute which served to check the rate of fall into the waters below would have the reverse effect and carry it away from the dropping place which had been chosen to create maximum disruption to enemy shipping. Fearing the attentions of the Royal Navy, German shipping tended to use routes which hugged the European coastline. Minelaying therefore had to be done in channels and at the mouth of harbours and estuaries which were usually close to land. Often mine laying aircraft would fly an area for up to half an hour while the navigator pinpointed the exact height and position for the drop, leaving the bomber vulnerable to attack by fighters and shore batteries.

Air gunnery training required aircraft towing target drogues behind them. Far from being a simple case of target towing while novice gunners took pot shots, training flights could be as hazardous as a bombing mission over enemy territory. A mistake in timing or communication between pilot and the crewman responsible for controlling the drogue could result in the steel tow cable whiplashing around the fuselage out of control and causing injury to anyone in its way.

Towing the funnel shaped drogues caused the heavy bombers to be even slower and more sluggish, presenting an attractive target to stray enemy fighters. During his wartime service one member of a target towing crew served in two planes shot down by enemy action, landing on a local beach and ditching in the sea. He also lost the tips of fingers in an accident while paying out the towing cable.

Airborne, all aircraft were liable to buffeting from air currents. Electrical storms in the atmosphere would turn the fuselage into an electrical conductor with the ever present risk of fire and all navigational instruments rendered inoperative. Adverse weather conditions when flying through cumulus cloud could cause condensation to freeze resulting in, at best, a drop in air speed caused by ice on the wings, or at worst, a loss of power when ice formed in an engine's carburettor.

When there were no missions arranged there were classes to attend on the ground, updating on new instruments, new techniques. Planes would be flown on air test to try out repairs or new components, or to train crews. Whole days were taken up, much to the chagrin of those living nearby, while planes flew in circuits around the airfield, barely touching the runway with the undercarriage before opening the throttle and taking off again. Often these bumps and circuits, as they were known by crews, formed the most concentrated daylight flying that local people ever saw. Many a mother had cause to rue Bomber Command's contribution to the war effort as plane after plane droned into the sky and circled over a house where baby had been put down for an afternoon nap.

6 A Place to Work

Whatever its importance to planners and militarists, the over-riding importance to those posted on an airfield was as a place of work. Air war was assuming a new importance but without an airfield's infrastructure of support and ancillary workers, the planes would not get airborne. And today, however much the remains of hangars, control towers and runways highlight the airfield's primary role, they give little indication of the variety and range of skills and trades centred around them which helped maintain the station as an effective fighting unit. And however much history books focus on aircrew and aces, the contribution made by other personnel on the airfield can easily be overlooked. In this and the following two chapters, consideration is given to some of the trades and callings working quietly and conscientiously behind the scenes; members of an airfield team who felt the fortunes of the squadrons they supported as keenly as any who flew with them.

Edgar Reeks was called up from his civilian job in a bank and assigned to the ranks of administrators. As an adjutant he was member of a commissioning party for a new airfield at Dunkeswell and was among the first servicemen to arrive and prepare it for the arrival of squadrons. His first task was to go to RAF Exeter and collect 50 men who were assembling there and proceed with them to Dunkeswell. At Exeter no one knew anything about Dunkeswell, much less the party of men due to transfer there. Edgar Reeks stayed on and remained vigilant and after a couple of days managed to round up about a dozen men and a huge pile of signals telling him why the others hadn't come: they were ill, on leave, overseas or had sick wives or children. Whatever, the reality was that the working party was short of manpower.

On arrival at the airfield site the adjutant and his depleted party were to discover there was no water laid to the buildings, no electricity, no telephone, no signals system, scarcely any transport, no stationery or typewriters and hardly any staff. In three weeks they would be expected to welcome three thousand Americans arriving in four squadrons. The English Commanding Officer who arrived shortly after Edgar was a peacetime career serviceman and could not comprehend the muddle that Dunkeswell presented. It was a nightmare for the commissioning team but they were successful in getting the station operational in time for the arrival of aircraft, but not before three weeks of intense work had cost Edgar 15 pounds in lost weight.

When a station became operational the role for administrators was only just beginning. Not only did planes and personnel need administrative support, systems and networks had to be set up for the station to function efficiently. Arthur Hart remembers failing the strict medical for aircrew and electing to become an Orderly Room Runner. This involved him collecting post from around the station headquarters and administration buildings ready for collection by the Motor Transport section who would deliver it to a central point for posting.

Gaining promotion to Aircraftsman I, his warrant officer showed him a list of trade vacancies which included among them an Alsation Guard Dog Handler. This appealed to Arthur, with memories of his parent's three year old Alsation at home. He signed up for training and was eventually posted to RAF Wroughton in Gloucestershire. When he arrived there the airfield was still new, covering more than one thousand acres and open for barely six months. Part of the airfield was occupied by a factory site and it was only later that he realised the gliders made there which he and his eight colleague handlers were guarding, were Horsa gliders being amassed to transport troops during the invasion of Europe.

Marjorie Catterall joined the Women's Auxilliary Air Force in 1940 and travelled to West Drayton where she was kitted out with full uniform and regulation underwear. She stayed there for six weeks during which time it was impressed upon her how to march, make perfect left and right turns, how to salute and to whom. Once a fortnight there was pay parade when she would receive 8/- (40 pence). When basic training was complete she was invited to choose a trade and was posted to Cranwell as a telephonist. There she lived in married quarters and was part of a small group of just 30 WAAFS on an airfield of 7000 men.

Her brother Harry Dobbing was an air-frame fitter earning rather more at half a crown (12½p) a day. After a childhood in Stockton-on-Tees where he had slept three in a bed, the accommodation shared with 20 men in a Nissen hut and individual beds was good by comparison. Lights out at 10pm ensured the tradesmen rising early in the right spirit to tackle their work of checking tyres, hydraulic fluid and checking the Wellington bombers based at Driffield for wing flaws. He worked with two engine fitters and after their inspection of each aircraft was completed, Form 700 would be signed confirming that all checks had been carried out – and Heaven help them if they had not.

* * * *

As a schoolteacher, H Grantham Hill was in a reserved occupation but he was keen and raring to serve his country. In summer 1940 derestriction took place and he applied for work as a meteorological officer. There were no vacancies at that time and so, in January 1941 he was sent to Blackpool for basic training as a wireless operator. When he could march sufficiently well and had mastered morse code at the rate of six words a minute, he was transferred to Compton Bassett, Wiltshire, to increase the action of his morse finger by three times.

Alas, at first he was not to put his skills in radio telegraphy to good use since on his first airfield he was made to work as a wireless mechanic. Like Harry Dobbing the air frame fitter, he was to carry out a daily inspection of wireless equipment in each aircraft and initial the book in the squadron flight office. Each aircraft had a page taken up in the large ledger and thus a central record could be maintained of all inspections, checks and work carried out on planes in the squadron.

Starting as Aircraftsman II he rose to the rank of corporal and qualified for 6d a day non ambition pay. He had no wish to rise through the ranks in his particular trade but still hoped to become a met. officer and find a setting where he could put the experience of his mathematics degree to better use. Eventually a vacancy for a Met. Observer occurred and Grantham Hill's life switched from wireless receivers to barometers, from signing daily inspection sheets to working out differential equations and plotting graphs to work out the curvature of isobars.

As an observer he was responsible for accumulating local data and forwarding it to headquarters at Bracknell where it would be incorporated with observations from adjoining areas and sent back in the form of an overall picture of weather conditions. With information on wind speed, cloud formations, humidity and barometric pressure, the senior meteorological officers on each station would interpret weather fronts, make forecasts and from them recommendations about the feasibility of flying operations. Grantham Hill remembers many arguments between Met. Officers and Station Commanders, one side concerned about adverse weather conditions en route to the proposed target and the other under pressure from officers at Group level to get their squadrons airborne.

* * * *

Edith Kup's first experiences as a servicewoman were as a driver at RAF Debden's Motor Transport section. Here she was expected to drive anything from a luxurious Humber Snipe staff car, to tractors and aircraft tugs, and Queen Mary's – the nickname for articulated airframe transporters. When a long drive was called for, two female MT drivers would be assigned: one to drive, one to map read and navigate on roads where signposts had been removed with the threat of invasion. When she was not employed in driving duties she could be enlisted to help out in the cookhouse until her MT Officer complained that his girls were not being employed appropriately.

On arrival at Debden after a day's journey from Leeds in pouring rain, Edith Kup was dismayed and of low spirits. When allocated to cold, damp quarters on the airfield she seriously began to question the wisdom of volunteering. But those feelings passed quickly when the billets were brightened with a few personal belongings from home. A bunch of flowers had been arranged in a vase on a locker and a tap with a plentiful supply of hot water was soon discovered.

After settling in she was introduced to the MT section and allocated a two and a half ton Albion lorry to drive. After a short trip with MT personnel on board, it was left to her to take the wheel. She recalls the truck being hard and heavy to control. There was no syncromesh on the gears and she remembers the bruise she received on her shin from a kick administered by her instructor when leaving her foot too long on the clutch pedal.

Following Dunkirk things began to change and she was moved from the MT section to become a plotter in the station operations room. Engaged in Radio Direction Finding she covered an area of the eastern sector of Britain. With a teller at headquarters who was in contact with radar stations, she would listen to broadcasts and plot the aircraft on large maps and diagrams spread out before her as they entered her sector. A controller sat on a balcony with full view of the plotting room, able to monitor the progress of a raid as planes passed from one plotter to another and from sector to sector. Working hours were four hours on, four hours off. If circumstances demanded it and a large raid was underway she would work an eight hour shift. Usually there would be a twelve hour period off each week to allow vital sleep to be caught up. By that time she had reached officer status. Standards of accommodation improved and there was even a batwoman to clean up and make the beds each day. As officers, she and her colleagues were allowed a drinks cupboard with an 'honesty book' to record drinks consumed and an expectation that debts would be settled without demur. But little time was spent in the billet despite its comforts. Watch keepers kept constant hours and there was always someone sleeping and in need of peace and quiet whatever the hour of day or night.

Eventually she was transferred back to Yorkshire, her home county, and posted to Pocklington. Orders arrived there by telephone tied line from Group headquarters. These would detail the route, target and bomb load. Once these were known the station's Commanding Officer joined senior staff in the Operations Room and all sections joined in the vital planning of the raid. Among crucial factors taken into account were aircraft serviceability – how many could be got into the air – and crew availability – injury or indisposition to one member could cause the whole crew to be stood down. Edith remembers those times of planning. There would be feverish activity everywhere, outside around dispersals where aircraft were made ready and in offices where final details were argued out. At such times the camp was virtually sealed with no private 'phone calls allowed in or out of the airfield.

Part of her work was to ensure that every crew member had their orders and that all flimsies, the rice paper sheets containing secret information and call signs, had been checked for accuracy. Details of captains and aircraft were passed to her for writing on the operations boards. It was at that point, as she wrote up their names in chalk on the wall board, that she began to have premonitions of which crews would return; premonitions which nine times out of ten would prove to be correct.

* * * *

One of the first things Mary Williamson did when she finished work as a bread roundsperson for the local Co-operative stores, and before commencing work on a local airfield as a mobile canteen driver, was to visit the local stationers and purchase a ready reckoner. Her job was as a civilian with the Navy, Army, Air Forces Institute (NAAFI) and would involve her in more than merely driving a two ton Ford mobile canteen around perimeter tracks to serve tea and cakes to waiting airmen. She would also be responsible for ordering the canteen's stock and for calculating and cutting it into sufficiently sized amounts which would make a profit. More than just a van driver, she would be a manageress in all but name.

She still possesses that ready reckoner and on the inside pages in her neat copperplate are recipes for drinks served. An eight gallon urn of ready made tea consisted of eight ounces of dry tea, thirty two ounces of sugar and two tins of unsweetened milk, or five gills if fresh milk was used. Coffee, which was less popular with English tastes, had a pinch of salt added and retailed at a ha'penny more than a cup of tea. Despite the bulk, ready mixed nature of the drink it was popular with airmen who hailed her visits as much for the break provided from the work in hand as for the quality of the refreshments purveyed.

Rock cakes and jam doughnuts baked at the NAAFI cookhouse on the airfield were also popular alongside pieces of fruit cake cut into bite sized pieces from a 7/- (35p) slab often obtained from a local bakery. When cake became dried and stale, wartime thrift and NAAFI lore dictated it should be chopped up, placed between pastry, baked and cut into squares and resold as Nelson's cake.

Her working day started at West Raynham airfield with a tour of the perimeter track, stopping at each dispersal point to open the van's rear stable-type half doors to create a servery. Alternatively, the sides of the van could be lifted to form a canopy. Cups of tea and cakes were served only to airmen, the expectation being that aircrew could find their comforts back at their mess.

From there she travelled to neighbouring airfields at Sculthorpe, Massingham and Foulsham – all within a fifteen mile radius of her home base in an area of North Norfolk densely populated by airfields and satellite sites. On her way she would also stop off to serve the crews of anti-aircraft batteries and searchlight emplacements. On her return to Raynham there was book keeping and stock ordering to complete – aspects of the job she remembers as being the least rewarding. And before the mobile canteen was garaged for the night it might need a trip to the station's MT section for refuelling in readiness for the next day.

Although a civilian, Mary Williamson was not spared any of the excitement of life on an airfield. On one occasion a fighter plane taxi-ing near her, clipped the van's roof with its wing tip and tore the canvas covering while she was serving from inside it. On another occasion she was parked on an apron near the technical site when a fighter crashed on the station generator, killing two people and causing an explosion which would leave her deaf in one ear for the remainder of her life.

Even away from airfields life could still be eventful. On one visit to a local market town to collect fresh produce for the NAAFI kitchens, her van's steering rod broke. With a woman's ingenuity she sought the only thing possible to make a repair: raffia from a bunch of flowers she was carrying. Making a temporary repair from this she proceeded to drive the four miles to the airfield where, upon arrival at the main gates, the raffia broke and her repair gave way. When personnel from the MT section arrived to tow away the damaged van they were amazed not so much by her ingenuity as her inability to see the dangers posed by driving a two ton truck held together by raffia!

When Americans arrived in the area and were spotted by Mary working on telephone cables at the roadside, she stopped the van and asked if they would like refreshment. Next thing she knew she had been lassoed by a rope from the top of the pole. However, when it was realised coffee would not flow from the urns until the manageress was properly installed behind her counter, the lasso was loosened, taken off and goodwill was restored.

For a short time men were also employed by the NAAFI as mobile canteen drivers. After working as a lorry driver delivering aggregates to airfield construction sites, Colin Acock became a canteen driver. With a small Ford 8 van at his command, life was a little more cramped than on Mary Williamson's two-tonner. The serving counter was slotted between the open rear doors and inside there was a working and food preparation area just two feet long. Somehow Colin Acock managed to fold his six foot body into this confined space, but frequently received blows to the head as he manœuvred with mugs of tea and sandwich 'wads'.

When he was transferred to RAF Coltishall he was given a canteen of improved design. Originally it had been an ambulance and with a height conversion even a six footer could stand and work in a measure of comfort. At the time, airfield defence was the responsibility of a contingent of soldiers from the Durham Light Infantry whose work was to staff the gun emplacements around the airfield. Part of his van round was to visit the outlying gun sites and Colin remembers that many of them would not eat their meagre chocolate ration, but instead collect enough of it to make a parcel and send it to families in the North East.

Left *Technicians and fitters working at a dispersal give a thorough overhaul to a Lancaster of 115 Squadron at Little Snoring airfield in Norfolk. A few weeks after this picture was taken it would overshoot the runway and, severely damaged, come to rest beside a local road (see page 61)* (K.B.M. Grant via T. Cushing)

Right *A salvage team prepare to lift a Hurricane of 17 Squadron with a damaged undercarriage prior to removal to the hangars for repair.* (Edith Kup)

Below *Armourers prepare to bomb up a 4000 pounder inscribed to Hitler from members of the Southgate Gang serving with 115 Squadron at East Wretham in Norfolk.* (Tom Cushing)

7 An Armourer's Story

When he passed out from training in May 1941, Jim Paxton felt a glow of pride posted to a squadron armoury when most of his fellow trainees were assigned to maintenance units, station armouries and bomb dumps. He knew that with bomber squadrons in the forefront of war he would be able to use skills learned during his long and arduous training. He would be working on the Blenheims of 107 squadron at Leuchars in his native Scotland, or so he thought reading directions posted on the bulletin board. He was certainly destined to join 107 squadron but not at Leuchars. On arrival at the station he was told the squadron were no longer there and it took a full 48 hours for administrators to discover its new location and send him and a fellow apprentice on their way. Many, many hours and hundreds of miles later they arrived in East Anglia, located the village of Great Massingham where their airfield was and reported for duty, only to be greeted by an officer informing them that they would be put on a charge for being absent without leave! Small wonder that his view of the unit and his new place of work was biased in those first weeks.

Jim was joining the squadron as a fitter armourer. The Royal Air Force had two basic armament trades, that of fitter armourer (group 1) and armourer (group 2). Additionally there were armourer's assistants and aircraft hands allocated to each armament section. Fitter armourers were the most highly skilled tradesmen in the armament field and were normally ex-aircraft apprentices or selected armourers who had undergone further extensive training at an armament school. Armourers were recruited from among aircraft hands or assistants, or by direct entry by personnel with a suitable educational and technical background. Armourer's assistants were general duties personnel employed in largely unskilled work and carrying out a 'mate' function working with skilled armourers.

As a trade, armament personnel on an airfield were responsible for servicing, maintaining and repairing all weaponry. This involved them working on a range of equipment from guns, bomb carriers to gun turrets, and involved fusing and loading of bombs. They had responsibilities also for the storage of all ammunition, explosives, fuses and detonators. The personnel were usually divided between the station armoury and the squadron armoury. Station armoury was responsible for station defence weapons, the bomb dump and storage of all explosive and pyrotechnic devices. Squadron armouries took charge of all matters relating to aircraft weapon systems, including maintaining guns and turrets, and taking control of bombing up and debombing the planes in their group.

With flint and brick cottages, church, chapels, shops and public houses arranged around a large village green broken by small roads, tracks and large ponds once attached to a long vanished abbey, Great Massingham looked every inch the archetypal English village. Yet only a few hundred yards from a memorial of Portland stone erected twenty two years earlier to mark the dead of another conflict, the rural idyll was to be shattered in fields to the east where a satellite station to West Raynham had grown among the gently rolling farmland.

Camp facilities were scattered around the village itself with the technical site on the fringe of the main village cluster and a cook house at its heart. Accommodation for most personnel was provided at the parent station but with their need to be 'on call' and working a shift system, squadron armoury personnel were billeted in a farmhouse on the opposite side of the airfield from the bomb dump. Mess facilities for officers and sergeants, transport and armoury sections, flying control and squadron flight offices were sited in the village or arranged around the flying field. In the early days it was a grass 'drome which churned up into a sea of mud during winter and wet weather. Wellington boots were normal footwear for many of the station trades and it was quite normal for them to turn up at a dance in the local village hall, leave their boots in the cloakroom and dance wearing thick woollen seaboot socks.

Life at Massingham for Jim Paxton was a mixture of working, eating, and sleeping, not always in that order or, in the case of the latter, with any great frequency. The squadron was engaged in daylight bombing, much of it targetted against enemy shipping off the Dutch and North German coasts. This meant Jim and his fellow armourers would be busiest in the evenings and at night time, engaged primarily on maintaining and repairing the Blenheim's gun turrets and assisting with bombing up aircraft prior to operations.

Regularly bombing up would be well under way, often nearing completion, when orders would arrive at the dispersal for a change of load. It was not unknown to have several changes in an evening and it was then common to call in the watch which had been stood down to assist those who had spent the evening manhandling the heavy bomb loads. A runner would be sent, not only around the station billets, but also around the village pubs to muster enough men to come and, literally, ease the load.

The much favoured but highly unofficial method of bombing up was to manhandle the load if 250 lb bombs were being used. The bomb would be lifted by nose and tail and lowered onto the bent back of the third team member. He would then stand up, slowly, lifting the bomb while colleagues eased the lug into the slip for fastening. After loading three or four aircraft with four 250 lb and four 40 lb bombs each a fair thirst would be worked up and it would be the task of the village pubs, rather than station mess, to provide the means to slake it.

Each section and flight had its own favoured drinking place among the village's three hostelries. At lunchtime the backroom of the New Inn would be home to men of the armoury section who would be drinking cider. So popular was it, that it acquired the nickname The Plumbers Arms after the armourers whose trade involved them in cleaning guns with a cleaning rod which resembled the process of rodding through a blocked toilet pipe! Later, patronage and loyalty switched to The Swan where the landlord's red haired teenage daughter was an added attraction to the airmen.

Everything on the airfield was geared to operational requirements and there were few parades. The only formal parades Jim remembers attending were Pay Parades and even they were flexible and irregular. With no fixed working hours social life centred on village pubs. Off duty time was spent letter writing, washing or repairing clothes. Once a day a NAAFI van toured the station bringing tea, sandwiches and the chance to talk with someone whose life was not involved in any way with the work of getting Blenheims heavy with bombs into the skies. The Women's Voluntary Service served the station similarly, with a small van laden with tea and small items of confectionery driven by a tall, elderly grey haired lady who provided another diversion from work.

For the armourers, transport to the home base at Raynham just a few miles away was not always easy and for much of the time Massingham was like living in a closed community, a world of their own dominated by heavy, manual work. Occasionally the whole squadron would be stood down and given a day, or part of a day, free of operational duties.

On these occasions squadron transport section would despatch a bus to the airfield and those who had money to spend or sufficient time off to allow for a trip, could fight for a place on the crowded vehicle and be driven to the county town. Once there, one particular hostelry was popular and often gutted, not by the Luftwaffe, but by RAF personnel engaged in regular inter-squadron battles. There was a high concentration of Blenheim bases in the area and squadron pride among members was high. But Jim Paxton was rarely involved in these pursuits. He was still receiving apprentices pay of 10/- (50p) a fortnight and in 1941 there were limits on how far a wage could be stretched.

For the young man from Edinburgh, three hundred miles away from home, life at Massingham was still tolerable and compared more than favourably with his experience of apprentice training school. Discipline there had been strictly enforced and was often harsh. Life at Massingham was more primitive and there were certainly fewer amenities than training school, but overall the atmosphere was more free and easy than before. Respect of senior officers and among colleagues was earned and there were rarely problems of discipline breakdown.

At first there had been a conflict between what he had been taught as an apprentice and what he saw in common practice among technicians on the unit. Arriving from training school he had been struck by their casual, almost cavalier attitude to the handling of bombs and high explosives. Yet within a month he was as bad, if not worse than the rest of his section. There were times, it would seem to him many years later, when his guardian angel must have been working overtime.

During his service at Massingham two new squadrons arrived for a brief sojourn. One squadron was flying Wellington bombers and the other was testing Flying Fortresses prior to their formal adoption by the RAF.

The personnel from the visiting squadrons travelled to and from Raynham each day and there was little contact between residents and newcomers, which considering squadron rivalry and unit pride was probably a blessing in disguise.

On one occasion the aircraft of 107 squadron took part in the 1000 bomber raids on Cologne, though for that particular operation they left Massingham to operate out of Horsham St Faith airfield near Norwich, where bombing up took place. Jim also remembers a trip to the parent station for a review of Bomber Command undertaken by Prime Minister Winston Churchill.

After a period of intense flying activities, in September 1941 107 squadron was sent to the Middle East and life on and around the airfield at Massingham became quiet. But the peace was to last only a few months. At the start of 1942 107 returned, rearmed with Boston III aircraft. A year later they moved on again and the airfield was treated to a facelift when runways were laid and accommodation for airmen and aircrew built. Massingham then became home to target towing and training flights until the arrival of a squadron flying Mosquito night fighters who remained at the station until hostilities ended.

After leaving the Blenheim squadron at Massingham, Jim Paxton served on an airfield occupied by a Free French unit. They too had a strong *esprit de corps* which also extended to any RAF personnel working alongside who cared to identify with them. Jim felt as much a part of them as his earlier "Blenheim brotherhood". The squadron had a RAF Commanding Officer, Adjutant, Flight Sergeant and orderly room staff, and a parallel French chain of command. The Maintenance Flight was composed of RAF personnel, ground crew a mixture of French and English, but aircrew entirely French. They are remembered by him as "a mixed bunch", drawn from a variety of sources including the French Air Force, Navy, Fleet Air Arm and Foreign Legion. He came to admire them as much as English squadrons, particularly when a Free French unit flying from Biggin Hill re-equipped from Mark 5 to Mark 9 Spitfires without withdrawing from operations or being stood down.

Jim also served on Lincolnshire airfields and was working as an armourer when a directive came down that aircraft must be debombed or their bomb loads defused in the event of ops being cancelled. This followed explosions at nearby Scampton when several Lancaster bombers were destroyed as they stood on the airfield with bomb loads intact after a mission had been scrapped.

After this Jim was ordered out to defuse the 4,000lb bombs already on board his squadron's aircraft after their takeoff had been cancelled. On arrival at the aircraft he found the detonator tongs were ineffective, so spying the section's WAAF driver nearby he asked if he could borrow a hairpin. Thus equipped Jim defused 'B' Flight, with the help of the WAAF who, on realising the predicament, assisted under the aircraft by gathering the detonators as they were removed. "It said a lot for her faith in the competence of the armourers she worked with, and quite a bit about her nerves! In view of what had happened so recently on the neighbouring station (Scampton), that took real guts in my opinion," he recalls.

8 A Defender's Story

When Freddie Forsdyke arrived at his first posting in Cambridge he could scarcely believe his good fortune. He was billeted in Trinity Hall, one of the university town's premier colleges, and took his meals in the college banqueting hall where there was waiter service. It was in contrast to his training but the luxurious lifestyle was only sampled for a few brief days. It was soon realised that Freddie and his colleagues were not trainee aircrew as first supposed. When it was discovered they were merely Aircraftsmen II, General Duties, things changed rapidly.

"When the penny dropped our life soon changed," he recalls. "They really went to town on us and we were soon made to learn every drill and rifle movement in the book." As though to punish the new recruits for the administrative error, the new intake were taught to march at 140 paces to the minute, when the normal RAF speed was 120 paces.

Freddie Forsdyke was called up on May 23, 1940, and ordered to report to RAF Uxbridge. By mid-day he had been issued with the mug and eating irons that would accompany him throughout his wartime service. That day he ate dinner in a hut where long trestle tables were layed out, covered by green oilcloths that were blistered and had clearly seen better days. Liver and bacon was on the menu (it was early in the war and bacon had still to become the food of the few), accompanied by potatoes and a particular type of green vegetable that seemed to Freddie to be seaweed from a distant beach. For dessert they were served rhubarb and custard. Thus fortified he approached his second day at Uxbridge, when equipment was issued to new recruits.

His equipment issue list was typical of all new recruits into the RAF and for much of their life on airfields these items would become the bulk of personal possessions that many had claim to. From the equipment store he collected: a kitbag; webbing; three pairs of socks; two pairs of boots; three pairs of underpants (which were so rough he was convinced they were made from a derivative of sandpaper); three vests with quarter length sleeves; three shirts; six collars; one tie; one pair of trousers; tunic and hat; shaving brush; clothes brush; two boot brushes, gym shoes; two pairs of shorts; two gym vests and a button stick. The last item on the list comprised a wooden strip with a slit down the centre, allowing it to be placed behind buttons about to be polished and so preventing cleaning materials staining the tunic.

New recruits were confined to Uxbridge for fourteen days. Freddie occupied his time stamping name and number with hammer and die onto brushes and identity tag. Between visits to the medical centre for vaccinations and check-ups he kept his uniform clean. There was a cinema on site and a NAAFI where for a penny-ha'penny he could purchase a cup of coffee which he says "resembled khaki emulsion paint and didn't taste much better." Billets were avoided during the day. Beds were convertible and folded into chairs when not accommodating sleepers. Mattresses were hard, there were no sheets only coarse woven blankets and a bolster pillow which was almost solid. Small wonder that those fleeting first days at Trinity Hall, Cambridge, where he shared an oak panelled room with three other recruits, dined off college plates and lived on the fat of the land, made such a long lasting impression on the young servicemen.

During his time at Cambridge his duties included guard duty at the local tramyard and at Trinity Hall where airgunners, navigators and wireless operators were undertaking their theoretical training. It was not long before he was posted across the border to the adjoining county of Norfolk and his first experience of airfield life at Watton. The station there had been among the first to open in 1939 and was a base for Blenheim bombers. Freddie arrived in a year which had seen heavy losses suffered by squadrons stationed there. An entire squadron had virtually been wiped out on a raid and only twelve planes had returned from a raid on Denmark. But the raids which resulted in those kind of losses were far removed from the work of aircraft hand Forsdyke, who on his first day there was despatched around the airfield to pick up every piece of litter in view.

Things hardly improved on successive days. His first work detail was planting cabbages on the airfield sports field. Joining two elderly civilians, one with a dibber to make the hole, the other with a seedling ready to drop into it, his first contribution to the war effort was to bring up the rear with a watering can to soak each plant. He then worked delivering coal to Nissen huts and quarters around the station, a hard and dirty job in which he needed three thorough baths before he could present himself at pay parade for his fortnightly wage of 5/- (25p). The majority of his money went on buying 'bull equipment' – boot blacking, razor blades, metal polish and dusters for button cleaning. Anything left over was zealously looked after and saved towards a weekend leave.

* * * *

In the early days of hostilities, airfield security was haphazard and often ineffective. Almost all were insecure, having no fences to divide them from the rest of the countryside, thereby providing easy access to any buildings on site. Some stations boasted pillboxes with gun positions and these were often sited at a point where a public road met the edge of the airfield and at which point travellers were turned back. At Christchurch in Hampshire the pillboxes were of advance design and retracted into the ground when not in use. For the majority of airfields, security consisted of dog patrols, rolls of barbed wire and ground gunners. However, things began to change when the threat of fifth columnists, saboteurs and enemy parachutists became real. The threat of enemy air attack added its own impetus to the need to secure airfields and to mount some form of defensive action in the event of strafing by enemy aircraft.

Still at Watton, Freddie became an Aircraftsman I and attended the station armoury on a course to familiarise himself with the Hispano Canon. The extent of airfield defence was shown in the words of the armourer who took the gun from its case and handed it to him with the words: "there it is and now you know as much about it as I do." A week later Freddie and the cannon were on the rifle range where he was allowed one round of live ammunition. Soon after he was detailed to take his place among the airfield defence party. At that time the airfield had five gun positions scattered over it. Most of the firing stock was old and small, only one of the gun emplacements being equipped with a Bofors gun.

For eighteen months his billet was a tent at the gun position near the main hangars. He worked there with six other men under direction from a corporal. Two men were on duty at any one time, one to man and fire the gun, the other to use the telephone handset to alert other gun positions and communicate with station headquarters. The gunsite was heavily sandbagged and part of its essential equipment was a pair of gallon flasks which would be filled up with cocoa each evening. Supper was taken to the gun crew at their position and eaten there, but all other meals were eaten in the station canteen. Each man had a rifle, of Canadian manufacture, which stood almost as tall as they did. It was their task to fire on any enemy raiders attacking the station and for this task the five gun positions and thirty or so rifles were to prove pitifully inadequate. Their efforts to bring down raiders, and those of similar detachments on airfields the length and breadth of the country, highlighted the need for an increase in the number of gun sites and rearmament with more effective weaponry.

A backup to the guns at Watton was the Parachute and Cable defence system, which was also in use at other stations. It utilised rockets to fire thin wire hawsers into the path of oncoming enemy aircraft as they commenced their attack approach. The wires would be held aloft for a short time by parachute before they drifted to the ground. The split second timing required to get the rockets in the air so wires could most effectively impede aircraft meant the system was not always wholly successful. Yet at Watton, thanks to the skill and timing of the spotter releasing the rockets, a notable success was achieved and Freddie Forsdyke was there to help clear up the mess.

On a cold February day in 1941 a lone Heinkel bomber managed to sneak over the English mainland undetected by radar. Fortunately a member of a gun team at Watton recognised the engine noise as different to that normally heard echoing from East Anglian skies. Thus alerted, he was ready to respond to as soon as the raider dipped from the clouds and began his approach. Quickly judging height, speed and angle the spotter selected six rockets from those positioned on concrete plinths around the airfield. His choice was correct and his timing in firing the rockets was perfect. Three cables penetrated the Heinkel's wings, bringing the plane belly flopping into a field close to a local village.

With the exception of a sprained ankle, the five man German crew were uninjured and when RAF servicemen arrived at the crash site they were busy around the fuselage trying to burn papers and parachutes before their captors arrived. The aircraft had its marking blacked out and the state of the wing fabric testified to the effect of the PAC system on that occasion, and to the skill of the operator who fired it.

The crew were taken to the local police station and airfield sickbay and waited until a party could be mustered to escort them to London for interrogation and eventual dispersal to a prisoner of war camp. Their mission had been one of armed reconnaissance over the Humberside area.

Freddie was a member of the escort party and was allowed ten hours away from normal duties to travel in a privately hired coach to London. It was an occasion for best blue helmets and gas masks, since the capital was in the heart of the danger zone. Boxes of sandwiches were packed and gallon vacuum flasks stowed away on board. The German airmen could have been forgiven for thinking that one particular group of British were about to indulge their traditional passion for picnics en-route! Only the composition and strength of the party suggested it was not the usual charabanc outing. Under command of an officer were five airmen with rifles, each with four bullets in the magazine and one in the 'spout' ready with the safety catch on. There was a driver, a sergeant and an interpreter who took up his position on the back seat of the coach guarding the injured man.

* * * *

Prisoner escort duty was a rare break in routine. After eighteen months in the hangar gun position he was sent to a position on the perimeter where a public road between Watton and Griston village crossed the airfield. There was now the threat of invasion and accompanying it a conscious tightening of security around airfields. Watton had boundaries marked by roads and hedges which could easily be patrolled and watched. The only vulnerable area was the road crossing the station so a white painted gate, sentry box, gun site and pillbox were established there.

Local people needing to use the road on a regular basis had an Air Ministry pass issued them. If a traveller arrived without such a document it was Freddie's job to 'phone his counterpart at the opposite end of the road and then escort the traveller through if permission was given. Where vehicles crossed in this way, the escort duty was undertaken using a 350cc Matchless motorbike. The motor cyclist always travelled behind any vehicle: 'Always let them lead' was the golden rule.

Each airfield had an armoured car assigned to it, as part of the new security arrangements. He recalls one memorable night duty in Watton's Alvis vehicle which was equipped with two Browning guns. It was the dead of night. The customary evening drink of cocoa ready mixed with sugar and Libby's milk had been consumed. He took up position in the armoured car closing and securing the hatch above him. Soon the cocoa was having an effect and the airfield defender drifted towards sleep. It would have been a restful sleep had it not been for a nightmare – not that it was particularly bad in itself but it had the effect of making him wake with a start. He immediately sat up, forgetting where he was and unaware in that split second that the hatch was shut. His tin hat made contact with the armoured plate and to a metallic ring the gallant defender slipped back into the seat and an even deeper, befuddled sleep brought on by a temporary state of unconsciousness!

Not every airfield had a purpose built armoured car. At Brize Norton the vehicle was improvised from a Bedford lorry with an armoured front and a large box section at the back. The sides of this rear section were one foot thick and filled with sand to prevent the passage of bullets or shrapnel. This provided a suitable nest for a machine gun crew and thus equipped the makeshift 'armoured' car would tour the airfield.

The need to improve airfield defence was eventually consolidated by the formation of the RAF Regiment, from which were created station defence flights and regimental squadrons. New kit was issued to Freddie and his colleagues, consisting this time of khaki battle dress and webbing, with army boots and RAF Regiment flashes for the tunic. Training in the use of the Bofors gun took place at Butlins holiday camp at Filey, Yorkshire, and he was then posted as a ground gunner to airfields in Norfolk, Hampshire and Devon, nowhere finding a billet which could quite compare with the University of Cambridge.

9 A Place to Live

Military airfields have always been more than just work places. They are home to the workforce, since to maintain an effective unit personnel have to be available quickly and easily. Modern trends are for those who wish, to live away from their base, with temporary accommodation provided on site when exercises or operations demand it. In 1939 all personnel were required to live on or very near their airfield and this required large areas of the site devoted to living accommodation.

Early on, accommodation was primitive and provided in tents. As installations grew to be more than a field where aircraft took off and landed, living quarters became more sophisticated with grandiose brick built barracks which took shape during the 1930s expansion period. Accommodation in such buildings was spacious compared with what would soon follow. They possessed central heating and offered enough space for individuals to maintain a degree of privacy. The exterior of the blocks was in mock Georgian style, built to a design which had received the approval of the Fine Arts Commission. For senior officers with families, estates of semi- and detached houses were constructed as part of the airfield's facilities.

But all this was destined to change in the rush to increase the RAFs fighting effectiveness when war broke out. No longer were barrack designs approved by aesthetic watchdogs, there just was not the time. Building methods and materials changed to those which would provide finished resources as quickly as possible. Timber hutments were bolted together on site, needing only a concrete base and plumbing. Nissen huts, made from sheets of corrugated metal shaped in semicircles, came in sections with windows or without and could be bolted together to whatever length the building's purpose required. Where brick was used it did not always present the warm red colour characterising buildings of 1930s construction. Now they were frequently covered on the exterior by a cement mixture. Whichever type of building was used they all shared the advantages of being easy to erect and quick to bring into commission.

Inside a typical billet, be it hutment or Nissen type, each side would be lined by beds. In the middle stood one, maybe two if the billet was particularly long, tortoise stoves with a tall black pipe rising from it to the roof to carry away smoke. There was no ceiling as such, only the curve of the corrugated sheets or the rafters and timbers supporting the hut roof. Each occupant of the residence possessed a floor mat at the foot of the bed on which to stand and dress. The floor itself was cold and nearly always shining and there would be a cupboard of cleaning materials at one end allowing the almost obsessional standards of cleanliness and tidiness to be maintained. A wooden locker and narrow wardrobe flanked each bed providing space to house kit and a few personal belongings and giving shape to the small space allotted to each man. In some billets there were additional shelves above the bed where three neatly folded blankets would be stacked for cold winter nights when the tortoise stove lost its glow.

Not only were stoves and floors kept scrupulously clean, everything that was part of the billet had its place and there were frequent inspections to remind errant airmen who forgot. Bed clothes had to be folded into a neat pile at the head of the bed, three blankets and two sheets layered alternately and bound overall in a fourth blanket which, by revealing its contents on two sides looked for all the world like some giant fancy cake. On the lockers at the corner nearest the bed would be cutlery in readiness for trips to the mess: knife, fork, spoon and pint mug. All other possessions were hidden away – neatly. Even the coke in the hods beside the stoves was often arranged to a neat peak and woebetide anyone spilling it while stoking up before an inspection. Many an officer's boot would descend on an offending piece of fuel and grind it into the floor surface before issuing the instruction to clear it up and get the place spotless again.

As adjutant, Edgar Reeks was involved in daily kit inspection and he remembers the ritual. "Every morning everyone stood by their beds. Blankets were all folded to within a fraction of an inch. Boots and kit would look absolutely spotless and they would all be laid out exactly the same. The presence of a piece of fluff on the floor gave everyone the shudders. Gas capes had to be folded absolutely flat and hung on the wall 'just so'. To ensure that they were really flat men used to get them properly folded and then iron them or sleep on them. They would be absolutely ruined for anti gas purposes but they would get good marks as they were completely flat! Overcoats had to be folded in a square fashion and be so precise that there was supposed to be a ridge along the edge. Actually, this was only obtained by inserting a thin piece of wood or ironing in a crease instead," he recalls.

In some huts there were two pictures pinned to the wall: one showing the kit layout for full dress, the other depicting ordinary everyday layout. If the inspecting officer or senior NCO found an item missing from an airman's kit there were rarely any questions asked. A 664B action chit would be completed so that the item could be renewed, the cost being deducted from the hapless airman's next pay. Clothing parades were sometimes more generous and an item could be replaced without charge if a senior officer was convinced it had worn through sufficiently.

Edgar Reeks also remembers visits to billets undertaken when an inspection had just been completed and the occupants were relaxing. "Some men would be reading, some writing and others would be playing cards. Some might even be doing a bit of carpentry. Others would be fixing bits of wire to the electrical supply to power wireless sets, irons or extra lights. The result of this was either a blaze of light – or no light at all if the fuse was blown because of an overload.

"Often bits of kit or lumps of wood would be drying in front of the fire, the latter to use on the fire next morning or to keep the present burn going. The occupants seemed to get used to the atmosphere of the billets and never seemed to open the windows or even draw the blackout – which needed constant chasing of all and sundry," he recollects.

In similar conditions the majority of a station's personnel – airmen and aircrew – would be housed. For those from poorer home backgrounds the utilitarian comforts of the billet were an improvement to life back home. For some, the sudden and enforced sharing of lives with nineteen or more others in cramped dormitories allowing no room for personal space or privacy, was a shock to a well-bred system yet one which they had no option but accept and come to terms with. As war progressed the camaraderie and sharing of life at its most raw was to heighten and boost the morale of men in the units in ways which those who planned the accommodation could never have dared hope. Allegiances ranged from their hut group, to their trade, from their squadron to their airfield. There was an overall pride and the communal life style of the airfield focussed an even greater loyalty. They knew only too well that they were all in it together, whatever their rank or status.

Of course not all airfields relied upon hutments and Nissen huts. One of the most unusual billets of all must have been at Newmarket Heath. The flat turf of the Rowley Mile horse racing course lent itself well to an airstrip and when the first squadron of Wellington bombers arrived from nearby Mildenhall in 1939 the crews and ground support staff were accommodated in the race course grandstand.

Bill Muddimer remembers the early days of the Rowley Mile airfield when facilities were primitive. Washing facilities consisted of a cold water tap and a large oversized bowl. Toilets were a six foot deep slit trench with hessian stretched around posts to afford a measure of privacy. There was an improvised seat consisting of two small posts and a cross piece which required users to sit forward to retain their balance. Users always set off to the latrine armed with a shovel, and after using the facility put a spade full of earth back into the trench.

Elsewhere town and country houses were taken over for senior personnel and transport was provided to take them to the airfield. The commanding officer of one station, well decorated with honours earned in battle and a regular visitor to the cockpits of his squadrons during missions, was known to live in fear of rats that prowled the rooms of his country house billet, causing him to always keep a handgun ready on his bedside table.

* * * *

There were other delights of rural life often reserved for the dependents of airman who had followed their husbands or lovers on postings by renting rooms or finding bed and breakfast accommodation in the nearest town or village. Ivy Armitage arrived with her husband when he was posted to RAF Cranwell as a clerk in the headquarters section. Air Force life had already given her a taste of country living during 1937 when they rented a thatched cottage at Henlow Camp in Bedfordshire but it did not prepare her for rural Lincolnshire.

She recalls the furnished cottage *sans* sink and with no piped water supply. "We had a well down the garden for drinking water and the usual rainwater butt outside the door which in summer had a thick green slime on top. To get drinking water we had to drop a bucket at the end of a long rope into the well. Very often when we pulled it back it would be covered in water snails.

"The lavatory was the thing I hated most. It was an ash pit type and had two seats. One was the usual height for adults and the other was lower for children to use. The farmer from whom we rented the cottage emptied it once a year and on that day I usually made sure I went to Lincoln."

She has memories of another rented cottage moved to after her daughter has been born during the Cranwell posting. One evening she was sitting with her small child in a large armchair. Suddenly she felt soft fur against her skin and heard the unmistakable sound of scurrying mice.

"They seemed to come from all corners of the room. I climbed onto the table, sat my daughter beside me and banged on the table with a shovel from the coal bucket to keep them away. My husband arrived home a couple of hours later and couldn't believe his eyes. He immediately went out to phone for a change of accommodation. We went to bed that night, the three of us in one bed. The mice were in the bedrooms too and my husband spent all night keeping them off the bed while we tried to sleep."

Compared to those experiences airfield billet with running water and toilet fixtures was luxury indeed. Even the constant coughing and snoring of fellow sleepers could not compare with verminous visitors.

As war progressed wives and fiancées of airmen were not encouraged to live near operational airfields. Too close a contact with their womenfolk was felt to put the men under unacceptable stress, placing them under undue pressure particularly when flying on operations. Some commanding officers positively discouraged relationships, insisting that wives only visit the station when officially invited to attend a squadron dance, and placing the responsibility on the women to arrange whatever accommodation they could for the night in the local village.

* * * *

At the first airfield she served on Edith Kup lived in fashionable airmen's married quarters. She was unmarried then and one of three single girls allocated to each house on the station. Even the occupiers of houses did not escape official inspections and there was the expectation that this billet would be as spotless as any other. Each day a sergeant from administration would brush a white napkin over surfaces to locate any tell-tale specks of dust. And if the cleaning and tidying had not been done to standard the whole task would be resumed and no one could leave until it was satisfactorily completed – annoying anyone with a limited time leave pass who was keen to be away at leisure.

Other postings saw her accommodated in huts on temporary airfield sites. Here it was often common for female ablutions to be in separate buildings, requiring her to make a trip across open ground, regardless of weather conditions, whenever the facilities of bath or toilet were required. On other occasions she stayed in local manor houses which had been converted into dormitory type accommodation, as many beds as possible being squeezed into the high ceilinged and formerly spacious rooms. Lists for bathing were posted to try and ration the limited supplies of hot water. For those not quick enough to get their name at the top of the list, who hated queuing and had sixpence to spare, they could visit a hotel in the nearest town and buy the facilities of a bath there.

The lack of obvious comfort in the billets wherever they were helped ensure that time spent in them was limited. Work took precedence and with opportunities for leisure provided on the station few spent longer than they needed resting in their huts. After sleep, a snatched game of cards and maybe the opportunity to write a letter home, the purpose of the billet was largely completed. There were too many other concerns and diversions for personnel to be put off by the quality of living accommodation. Whether it was a hut of wood or metal, married quarters or manor house, it served merely as a base where a few belongings could be kept, a place where a busy day or night would end and another begin.

10 A Place to Play

The old adage "All work and no play makes Jack a dull boy" has always been taken to heart by the Royal Air Force, never more so than during 1939–1945 when it recognised that the stresses and demands which war placed on servicemen should be balanced by opportunities for recreation and leisure activities. The occasional forty eight hour pass gave the opportunity to travel home to wives, girl friends or families or to take a trip to the nearest town offering pubs and clubs but there was still a need for leisure time outlets above that provided by messes and many airfields had cinema huts which doubled as theatres.

A sporting force was a fit force and there were plenty of opportunities for organised games with many an airfield having its own football pitch set on the acres of grass between the runways. Sport placed emphasis on teamsmanship and healthy rivalry and to further its pursuit some stations had squash courts or maycrete huts kitted out as gymnasiums. Not all sport was active: in places exhibitions provided high spots in the airfield sporting and social calendar. At Honington a boxing display given by a sergeant on the base received a good reception. Heartened by this success the station organised an All In Wrestling Show staged by ENSA but this did not prove at all popular.

Not all personnel were sporting and sought diversions away from the inevitable card games and letters home. One popular activity was making models of aircraft, 'scratch built' from whatever scraps could be easily and cheaply obtained on the station. They ranged in size and complexity from a scaled down Lancaster bomber made from wood with a wingspan as wide as an arm's reach, to Spitfire fighters formed from a beaten and flattened penny piece cut to shape, soldered onto a safety pin and offered perhaps to a special lady friend to wear as a brooch.

There was always someone on the station proficient with a paintbrush and many a bleak brick wall was enlivened by a large mural, often of an idyllic country scene bringing its own peace to the men whose home the huts were. Scenes also depicted aircraft from the squadron. On American stations paintings of scantily clad women were popular and were often sited near an exit door and used as a good luck charm for departing aircrew who remembered to plant a kiss on whatever part of the female form was represented.

Dances were organised and mess halls would be transformed by flags and bunting and a revolving ball to reflect and scatter light around the impromptu dance floor. At one end would be positioned the station band or an ensemble of servicemen drawn from across the ranks, playing whatever instruments they could muster. Occasionally well known bands would be engaged to play. Ted Heath and the Squadronnaires visited many airfields and Glenn Miller visited bases where American forces were stationed, playing concerts to vast crowds in hangars taken over for the night to be used as concert halls. On those occasions transport brought in personnel from fields in the neighbourhood and so full would the hangars be that it was not unusual to find men had climbed the metal support structure to obtain a "view from the gods." Not all visiting bands had a name and reputation as evergreen as Ted Heath and Glenn Miller and one service wife who went to a dance played by Sid Millward and his Nit Wits, still wonders whatever became of him!

Every Sunday at Debden in Essex there took place an invitation game of housey-housey in the Sergeants Mess, with dancing to follow. At Pocklington in Yorkshire officers got together to provide entertainment for their fellows in a dramatic production. Unfortunately the performance of Thunder Rock never progressed further than the first act, interrupted by the audience being called to carry out an emergency operation against the German warship Sharnhorst which had been spotted in the North Sea. Ninety two heavy bombers swept a wide area of sea in search of the ship and its accompanying flotilla but fog that night was to prevent any sightings. That was April 1940 and as war progressed the officers never found themselves with time enough to rehearse plays and that first act was the last piece of home produced dramatic entertainment staged there. Thereafter drama enough could be found all around them.

When there wasn't a dance taking place in a station mess, there could usually be reckoned to be a 'hop' of some kind taking place in a village hall in the neighbourhood. At Cranwell the village hall was the venue for regular dances, arranged without refreshments of any kind and with RAF personnel frequently outnumbering civilians. The female population was always in great demand to dance. One local girl recalls the expertise on the dance floor of lads from Glasgow who always excused a great deal during the gentlemen's Excuse Me. And at every dance, whether in mess or village hall, the last waltz of the evening was inevitably a slow, emotive rendering of "Who's Taking You Home Tonight?".

Passes were issued for 24 and 48 hour leaves. Public transport and the railway service in particular was subject to delay and difficulty. Often a trip which in peacetime would take a few hours would be doubled or trebled in time while trains stopped during air-raids, took circuitous routes where sections of track had been destroyed, or gave precedence to troop and military traffic. For many it was difficult to travel home in the time allowed so the next best place to spend the leave was a large town. The cities of Norwich, Ipswich, Cambridge, York and Lincoln all attracted servicemen from airfields in the surrounding countryside.

Lincoln High Street was a sea of Air Force blue in those days and there was always a queue at the Ritz Cinema's café despite many occasions when the menu was only able to offer tea and baked beans on toast. The city's Theatre Royal put on variety shows popular with airmen and aircrew alike. Variety shows toured airfields organised by ENSA but there was a feeling widely held that many of the acts were second rate, the best having been sent off abroad to maintain the morale of troops far from home. Dances at Lincoln's Drill Hall and Assembly Rooms were popular and one of the longest wartime queues at the Regal Cinema in the High Street was for a showing of Dangerous Moonlight, a film about the RAF in wartime.

Many airfields organised 'Liberty Runs' to local towns using crew buses or canvas tilted trucks, but not all servicemen had the inclination or finances to go.
 For Robert Tumulty, with home and family in Northern Ireland, posted to an

airfield in the depths of rural Norfolk, five miles from the nearest village sporting a public house, leave periods were spent helping local farmers at work in the fields. The Women's Land Army had still to penetrate the remoter country areas and there was an urgent need for extra labour to help in the country's food production. The chance to plant or pick potatoes and hoe turnips and sugar beet provided a welcome fillip to the income of a Leading Aircraftsman on a fortnightly pay cycle. And there was also the possibility of a boost to the diet when invited to the farmhouse for a breakfast of bacon and egg – a rationed luxury restricted on the airfields to aircrew returning from missions.

Robert Tumulty remembers to this day the kindness of local people who invited him and his countrymen into their homes, knowing well the isolation they felt, and the impossibility of returning home on even a 48 hour pass. One incident stands in his memory of handing in a purse at the local police station after finding it in Norwich. Leaving his name, number and station he left the police station and thought little more about it. Soon after he was surprised to receive a postal order for 2/6d and an invitation from the owner of the purse to spend his next 48 hour leave as a guest in their home. At that time the reward alone represented half a week's wages and Robert thought he was rich.

But at High Ercall in Shropshire there were no leaves unless they were earned. Station regulations laid down that everyone would have one day off each week but during the day of rest all ranks would be expected to do four hours gardening. A forty-eight hour pass would be issued once a month, but in order to qualify, a chit had to be produced showing that eight hours gardening had been completed. All ranks had to apply to the sergeant in charge for a printed card confirming their labour and without it the adjutant would not be able to issue the vital pass to 48 hours of freedom. Needless to say, things changed when the Commanding Officer was posted elsewhere. When the Squadron Commander took temporary control of the airfield in the meantime, morale and the standard of food served in the mess halls both rose dramatically.

At Dunkeswell there was only one place of amusement and that was the public house in the village around which the camp had been built. Beer supplies were frequently exhausted when a large group of fliers arrived so an enterprising landlord made up for deficiencies in ale by offering a rough cider. Former station adjutant Edgar Reeks remembers the real attraction of the pub being housed in a yard at the back: a pig. Most of the customers took delight in trying to get the hapless creature drunk and the animal seemed to have a capacity to stow away coarse cider that could rival any of the most hardened drinkers from the camp. Eventually the landlord had the pig killed but it was not until the porker had enjoyed many drunken nights scarcely able to stand up in his sty!

In an effort to provide suitable entertainment for the men of Dunkeswell, a committee of local ladies was set up to arrange dances in the nearby town of Honiton. Men would take the floor one night, NCO's the next and officers were relegated to just one night a week. Dances were also held at the station and when the Americans were established there they sent buses all over the locality to bring in local girls. They became a tradition and many a young lass in Taunton looked to the airfield to brighten the routines of wartime life. On one occasion though, the Americans were too preoccupied by the bar and its contents, totally ignoring the girls that had been bussed in. It was not long before the indignant females were approaching the Commanding Officer to get them home – but not before they had consumed their share of alcohol.

As adjutant it fell to Edgar Reeks to take responsibility for the Sergeants Mess which had a membership of 800. As officer in charge he was custodian of their funds and also of their private worries. He was even able to "pull a few strings" by contact with other NCOs in an attempt to settle some of those concerns. The bar was a thriving concern and during one four month period made a profit of £875. Polish squadrons were traditionally heavy whisky drinkers and on occasions when a Polish squadron passed through, bar returns could be relied upon to be high.

Problems for Edgar came after squadron 'flings' and on one occasion a group smashed more than 150 pint tumblers, carried the furniture into a heap in the middle of the floor and doused it with the contents of fire buckets and extinguishers. He recalls the Squadron Commander thought him hard when requesting that the men responsible pay for the broken glasses but when he was shown the extent of the damage the Commander organised a party of twenty five aircrew (not airmen) to scrub the place out, supervised by a squadron leader from the offending squadron. The year he was OIC, Sergeants Mess, a barrel of beer went missing on Christmas Eve followed by a side of bacon on Boxing Day. On New Year's Eve two more barrels of beer went absent without leave and the empties were eventually discovered floating in the local river. Nobody ever owned up.

* * * *

At the outbreak of war an organisation was formed to provide professional entertainers for the forces and war workers. That organisation rejoiced under the title Entertainment National Services Association, ENSA for short and Every Night Something Awful by popularly held belief. In 1942 the organisation staged an entertainment on the steps of St Pauls cathedral in London with celebrated actors and actresses taking part with choirs and military bands conducted by Sir Henry Wood. Less grand in scale and content, were the ENSA concert parties conducted around the countryside to airfields and other installations by Len Tostevin in his converted motor coach.

At the time he was an excursion coach driver working for Southdown Transport out of Brighton. On the first of November, 1942, he moved across with his vehicle to the service of ENSA. Back seats were removed from his up-to-date Leyland coach to create storage space for scenery and hampers, while the 20 remaining seats were available to the artists and their travelling pianist who provided sole musical accompaniment. Len Tostevin would meet each concert party at a convenient local railway station, taking them on a tour of local installations lasting between two and three weeks. At the end of that time they would be taken to the rail station for the next stage of their tour, and Len would collect the new group and show them to their billets which would be home for the length of their stay in the area.

Shows with titles like The Caledonians, All For Fun, Mirth and Melody gave an indication of the content, if not the quality. Straight drama was popular with Hobson's Choice and Chocolate Soldier being received well by service audiences. But, however good or bad, variety always went down well on the airfields. Audiences would be generous with their applause for groups and turns they respected but were quick to heckle, stamp and shake the hut's walls for those they did not. Acts were not always well known: Billy and Connie Faka, a comedy duo,

***Above** A good sense of humour was essential to survive on an airfield. Here the official crest of 192 Squadron has been repainted, seaside postcard style, by an airfield wit.* (A.J. Berry)

***Left** When there was no flying, training or maintenance to be done, personnel often gathered to pass an idle hour in a card school.* (Edith Kup)

***Below** Prangs were an occupational hazard in the air and on the ground. Here at RAF Debden a visiting Spitfire from 266 Squadron has collided with one of 17 Squadron's Hurricanes.* (Edith Kup)

Mario and Nesta, a song and dance act, and the comedian Billy Merson have not lived on in the annals of show business history. Concerts of classical music also toured and many a pianist and violinist died a death on the platform of a Nissen hut in some far flung corner of England.

On arrival at an airfield, Len would be greeted by a fatigues party standing by to help unload the coach. The ENSA entertainments manager travelling with the concert party was authorised to pay 1/6d (7½p) to all those who helped and there was never a shortage of willing hands from among station personnel. Airmen would be enlisted to tear tickets in half as patrons entered the hall. Front of house would be the responsibility of the ENSA entertainments manager who was powerless to quell the customary chorus of "Why Are We Waiting?" as the concert party prepared themselves behind the curtain or made last minute adjustments to their costume in the coach outside.

However things did not always run according to plan and life for a coach driver on wartime airfields was not without its dangers. On one occasion when the concert party coach arrived at Bexwell aerodrome, the Commanding Officer cancelled the show and ordered all personnel outside to clear the runway of snow and ice so that flying could take place. Thus a wandering troupe of minstrels, jesters and their driver found themselves chipping away at hard packed ice and grateful for the RAF tea and coffee which flowed to keep their hands and spirits warm.

During a visit to a Suffolk airfield Len was driving his coach across the 'drome to position it near the hall where the night's performance had taken place. It was after midnight and the scenery and clothes hampers had to be loaded up ready for the next call. The airfield's squadrons had been stood down and there had been a full house to enjoy the show earlier than evening. No flying was taking place and there was an eerie quiet about the airfield which the coach engine only barely disturbed. The driver, like all good coachmen, was thinking about the route to the next venue when suddenly lights came on all around him, lighting up the runway he was crossing with a glare like daylight. Then came the roar of Rolls Royce Merlin engines and a dark shadow passed overhead. Instinctively he slammed his foot on the accelerator pedal and as he did so the lights faded and the sound was lost. Ten days later on his next visit to the station he discovered than an aircraft from a West Country base had got into difficulties and requested permission to land at the nearest available airfield. Station control had given permission readily enough and switched on the runway lights and flarepath with scarce a thought for any stray Brighton buses which might be crossing their East Anglian runways at three o'clock in the morning!

11 Places of Danger

As soon as construction work on an airfield was underway it became fair target for enemy aircraft, bringing the wrath of war deep into a rural England which might otherwise have escaped attack and bombings.

Interruptions to work caused by enemy attacks were commonplace in the working routine of Peter Cron as he laboured on airfield sites. Bolt holes and rough shelters were provided around the area and there were times when an enemy aircraft overhead gave the opportunity for an unscheduled break and a quick hand of cards in the shelter while awaiting the all clear. There might even be the chance of bonus payments if there were any bomb craters to fill in. The possibility of enemy attack deep in the heart of the countryside did not seem out of place to Peter Cron who believed that if you worked on a site which would play an important part in the war effort it was reasonable to expect the enemy to do its utmost to hinder progress.

Less reasonable though was the assumption by one Luftwaffe pilot that three labourers driving home from the construction site at a sedate 30 mph through Norfolk lanes should be considered fair game for attack! Yet that was the fate of Peter Cron who had begged a lift home with a local publican who spent daytime 'moonlighting' on the construction site. It was a lucrative sideline at a time when pub takings were down and had helped him afford the brand new Ford car in which he drove Peter and another worker.

The car still had the smell of new leather and the shining, polished black bodywork attested to the fact that it was the publican's pride and joy. Suddenly the peace was shattered by the noise of an aircraft swooping low at them from the clouds. Dust and granite chippings rose from the road surface as a line of bullets strafed the car. Instinctively the occupants dived down beneath the dashboard and behind the front seats. As the driver did so he lost control of the steering wheel and the vehicle careered off the road. The German pilot did not return. He had been returning from a raid on the airfield site and seeing a moving target on the road below decided to loose off a few rounds and terrorise the locals before returning across the channel. The only injuries he caused were to the driver who, in those pre-seat belt days, had been ejected into a holly bush. The scratches on his face and neck were nothing compared to the scratched and dented panels on his once pristine car.

Raids on airfields increased, both in frequency and firepower, as war progressed. At Coltishall six bombs were dropped on an unfinished hangar and at Feltwell the main runway was pitted by 22 craters before a second force flew in dropping incendiaries. At Honington John Wortley remembers one of the hangars being hit by a bomb and put out of use. Intensive efforts followed to restore the building and four months later at nine in the morning the builders moved out, the new hangar was handed over and commissioned immediately. At four thirty that same day enemy raiders returned and scored another hit on the building.

The farmer also remembers one German pilot displaying particular daring and cheek by regularly raiding the airfield. So predictable did his visits become he was nicknamed The Honington Dornier. His plane usually followed the same course: along the Blackbourne river, banking right at Fakenham Magna church then low over Broom Hill woods. He was at an altitude too low for the station's outdated naval Bofors guns and he used his advantage to machine gun anything on the station that moved, and quite a few objects that didn't.

Unsubstantiated stories tell of a young soldier based at the airfield whose parents had been killed in an air raid. He was determined to avenge their death in some way and he took up position to await The Honington Dornier's next visit. When it appeared he ignored instructions not to shoot unless ordered and opened fire when the aircraft was in range. With an amazingly accurate opening volley he fatally wounded the pilot and the plane crashed in flames on the apron in front of the hangars.

A popular tactic of the Luftwaffe was for their planes to shadow returning RAF planes as they returned to their airfields. In this way they would enter English airspace unchallenged, their radar 'blips' hard to distinguish among the returning Allied forces. The intruders would then select the weakest of the returning bombers, those already damaged by fighters or anti-aircraft fire, and proceed to attack them before they could reach the safety of their airfields. Often the attacks occurred over the airfield. When this happened chaos ensued on the ground where planes that had landed successfully could still be written off by a newly shot up bomber landing out of control, or as often happened, by a plane falling out of the sky.

Part of Bill Muddimer's service was at an airfield where cadets undertook flying lessons. The field was of temporary construction and in use day and night. During the hours of darkness a flare path was laid to mark the landing strip and he was one of the team assigned to setting out the flares at regular intervals.

"The flares were difficult to put out and one night German planes were coming over and we had to run like hell to put them out. One flare was extremely hard to get rid of and the only way we could get rid of it was to stand around and water it in the natural way. It did the trick and with our airfield in darkness the planes bombed Brize Norton instead," says Bill.

No matter how diligent personnel were in checking lights and flare paths when enemy raiders were overhead, their best efforts were sometimes let down by revellers on the domestic site. Colin Acock had been enlisted to serve behind the bar for a party in the airmen's mess.

"The atmosphere was heavy," he recalls. "The beer was warm and so was everyone in the room. Someone decided to open a window.

"The curtains parted slightly, un-noticed by us but not by an approaching plane which was not one of ours. Suddenly all hell broke loose. Our Bofors guns opened up and the dark sky outside was lit up by the red tracers. Outside the anti-aircraft gunners could see nothing."

The German aircraft had vanished as quickly as it had come, but not before it had released a string of five bombs. The first landed just short of the block where the party was taking place, the other four dropped on the flying field.

"As it was I couldn't drive my NAAFI mobile canteen out of the yard because of

Top A Lancaster II *from 115 Squadron overshoots onto the road at Little Snoring in Norfolk. Only weeks before the aircraft had received a major overhaul (See page 40)* (K.B.M. Grant via Tom Cushing)

***Above** A German Ju88 brought down after attacking a nearby airfield.* (Edith Kup)

***Right** One way of coping with pressures and emotion of war was to find a quiet corner of the billet, bring out treasured pictures of family members, and write a letter home.* (Tom Cushing)

***Below** Aircrew of 199 Squadron pose in front of a Halifax bomber. Add a similar number of personnel for ground crew, multiply by the number of squadrons resident and add a liberal number of administrators and civilians to appreciate the population of airfields. Small wonder they frequently dwarfed the towns and villages nearby!* (A.J. Berry)

a hole in the road so instead of a nice warm bed in my digs at the village I spent the rest of that night on chairs in the canteen. It wasn't until morning that we finally saw just how lucky we really had been. If those bombs had been heavier and released later I certainly wouldn't have been around to serve any more NAAFI char," he says.

* * * *

Danger did not come solely from enemy attacks. With so many aircraft in the skies and on the ground, with servicemen working and flying long hours, human error inevitably played a part in accidents on stations. Incidents ranged from damage to property, like the taxi-ing plane which clipped Mary Williamson's NAAFI canteen, to loss of life, as in the case of groundcrew who forgot for one fatal moment that aircraft propellers were spinning and stepped into their lethal path.

Many an adjutant found himself preparing detailed reports on accidents ready for the internal boards of inquiry, staffed by senior officers. Among the thousands of incidents investigated were the Mustang aircraft which succeeded in demolishing the control tower at Bodney and the Stirling which did not gain enough height taking off from Newmarket Heath and crashed with mines on board and a full load of fuel. At Binbrook, Lincolnshire, an electrical short circuit in a Lancaster caused the entire bomb load to be discharged over the airfield. Among the destructive power raining down were incendiaries and two 500 lb high explosive bombs. Blast and fire damage was extensive and three planes were damaged on the ground. Rapid repairs took place to the runway and that same evening eighteen aircraft were able to take off for a raid on Cologne.

The Americans were not spared their share of accidents and at Thorpe Abbotts a gun in the ball turret of a B17 aircraft went off as the gunner was removing it for cleaning. The plane was on the ground having returned from a raid on Munich. As the machine gun fired its bullets the turret began to revolve, peppering the surrounding area with bullets as it did so. Servicemen in the vicinity ran for cover, including the gunner who, in running away from the aircraft, was mown down and killed by his own bullets. Three aircraft were damaged and one bullet punctured the wall of a farmhouse which stood almost a mile away from the dispersal where the incident occurred.

Every airfield knew accidents and catastrophes. When disaster struck, in skies at home or abroad or on the airfield itself, the usual method of coping was simply not to talk about it. There was rarely any outward sign of grief and often the only acknowledgement there had been a loss of life would be with those who had witnessed the incident. John Wortley visited Honington soon after two Wellingtons had collided. There was no talk of the accident on the station, not even rumours, he remembers. Had he not seen the accident with his own eyes he would not have believed it had happened. Life at Honington continued in the same way. Nobody said a word, preferring to keep their inner thoughts and feelings private. On the surface at least, life went on as though nothing had happened. It was as though the biggest risk on an airfield lay in simply acknowledging the very existence of danger.

12 Coping with Emotion

Rural wartime airfields were concerned with the business of death and it was not always possible for death and dying to be restricted to enemy skies. For members of aircrew, death was never far away, whether they were on a mission or stood down at the airfield. Even when danger was not taunting them, there was the ever present memory of colleagues who had not returned from sorties, the image of planes disintegrating over the target as flak hit home and the knowledge that any day the same thing could happen to them.

Under these conditions men developed their own mechanisms to cope with the range of feelings and emotions caused by their service. For some, the squadron itself provided the support they needed, offering the fellowship of an exclusive club where members shared the same risks. It was a support without words demanding in return a loyalty and devotion which themselves were armour against the worst which fate might hand out.

Others had a religious faith which was acknowledged by the presence on airfields of chaplains for all main denominations and buildings set aside for worship. Sometimes the gap between serving men and their chaplains, who were vested automatically with the rank of officer, was too great to bridge and on some airfields the main virtue of 'sky pilots' was a support in making applications for compassionate leave. Many Air Force chaplains themselves felt at a disadvantage since they could not visit airmen in their barracks in the way a civilian minister could visit parishioners in their homes.

There were churches and chapels in nearby towns and villages but occasionally they lacked a welcome for servicemen. During war service Bruce Blackie visited many local churches during his postings and found the buildings damp, cold and musty. Worse, members of the congregations afforded the military little in the way of a welcome, seeming wary of the presence of blue uniforms and perhaps fearful for the chastity of their daughters when the religious service was over! Further, many servicemen felt that civilian ministers had little idea of their particular spiritual needs and no knowledge or understanding of the conditions in which they lived and worked, or how their spirituality was affected by it.

"As I moved around different aerodromes," recalls Bruce Blackie, "I discovered that, in general, non-conformist chapels were well lit, warm, inviting and friendly places. The members sang lively, energetic hymns and showed uncommon hospitality to airmen with cups of tea after the service and invitations into their homes.

"This ungrudging welcome into their homes must have cost these civilians some of their meagre rations, yet by inviting us in they gave us the feelings of home we missed so much."

One of his postings was to a fighter station in the New Forest, near Ringwood. The local inhabitants had, as in many other places, accepted the invasion of their privacy with good grace, even though their property and lives were at risk every time enemy fighters attacked the station. Airmen were housed in the usual Nissen huts, furnished only with iron bedsteads and blankets. Even the customary thin RAF mattresses known as biscuits were missing. When Bruce wrote to his mother telling of this and using the RAF vernacular term for mattresses, she used a mother's love and considerable ration coupons to obtain a packet of sweet biscuits to parcel up for her son! It was after settling into these spartan billets that he came across meetings of the Soldiers and Airmens Scripture Leaders Association. Meetings were held in a small non-conformist chapel in the village nearby, using the chapel's Sunday school room. The meetings were held on Wednesday evenings and attracted civilians from the village to attend. One local person even wrote a letter to the airmen telling how inspired his fellow chapel members had been that young men in uniform should unashamedly stand up for their Christian faith. Links between airmen and villagers were soon built on these foundations and many an invitation was offered for servicemen to leave the tortoise stove in the Nissen hut and share the warmth of a cottage fireside.

"The bond between the airmen at the meeting and members of the chapel subordinated any ill feelings the chapel might have harboured against the RAF, especially when the white fence around their little cemetery suffered damage from airmen a little too exuberant after imbibing rather freely at the local inn," he recalls.

There was another active branch of the Soldiers and Airmens Scripture Leaders Association at the emergency landing ground at Manston in Kent where Bruce also served. He was one of many people who found that guidance and teaching of Scripture Leaders, who were airmen like themselves, communicated directly with their own experiences, without the need to bridge the gaps of rank. It was a practical approach to religion, meeting a need which civilian churches and military chaplains could not do as effectively. Significantly, on demobilisation many former Scripture Leaders left the service to become ministers.

A strong Christian faith was important to Bruce Blackie in coping with the range of emotions that conditions of war engendered. When he entered the RAF he reckoned his faith was 'immature' and found that service life and a home on a wartime airfield would challenge his beliefs more than once.

"Despite the losses of the Great War, it seemed fashionable to sweep the fact of death under the carpet, almost in the way the Victorians ignored the problems of sex. But when the chips were down, what the Bible had to say about being with Christ was not pie in the sky, but the anchor of inward peace and serene hope," he says now.

* * * *

Many a town with an airfield close by became witness to a funeral party passing on its way to the local church and cemetery. For Edith Kup as a member of the WAAFs there were many occasions when she was called to take part in funeral parades and, forty years on, she is still haunted by memories whenever The Last Post is played.

"The crews were all known to us. One became involved in the battle because we knew all the pilots. We danced with them, went out with them, laughed and joked with them. We could hear the battles because the RT was open in the Operations Room," she recalls. "Each battle was an agony until they were all safely down again and those who were shot down doubled that agony until we knew whether they were safe or not. When we lost one it was like losing a member of the family."

She has vivid memories of those funeral parades. "There would be a band and firing party, slow marching to the local churchyard. There would be a volley of rifle fire into the air and then the most awful, individual approach to the foot of the open grave by each officer who would salute. We were marched back to the camp in double quick time. That night there would be drinking in the mess. As WAAFs we went to a local pub for a rather sombre drink. We simply could not mourn all of our 'families' in the accepted way. Grief was experienced privately. One did not cry in public. I'm sure the general public did not understand and must have thought us uncaring."

As one of a complement of women on an airfield who were subject to the same service rules and discipline, a particular affinity developed with the male aircrew. It was an unspoken bond, deepened for those in the Ops Room by hearing the crew's radio broadcasts while raids took place. As a technical officer in the plotting room, Edith Kup had a special, emotional role to fulfil additional to that required by her uniform.

"We saw the crews take off in summer time, watching from the control tower roof. Then we would see them return for their debriefing, cold, grey and frightened before they could put their 'masks' on. They knew we could never give away confidences. Often they would catch us when alone and virtually cry on our shoulders. All the fear and tension would flood out, and that included the fear of showing fear," she recalls.

There were other occasions, and they were many, when she would accompany an airman after debriefing on a walk around local lanes. She would listen to confidences, give advice and comfort where at all possible. It was a relationship of mother to son, of sister to brother. Inevitably it could be a relationship liable to misunderstanding by WAAFs who served in administrative posts.

"No one ever gave way though. We were all aware that we each had our 'sons'. Because of this bond we were often hailed in the mess to share a drink and be treated as equal members of the squadron," she adds.

* * * *

Public houses in adjoining towns and villages were popular places for airmen and aircrew in off duty hours, and many local hostelries became the focus for squadron rites and frolics. In some public houses on the night following a bombing raid a spare pint of beer was always purchased, placed on the bar counter and left there in silent tribute to those who did not return. At the end of the evening as closing time neared, a barman or barmaid would consume the drink while the crews stood in witness to colleagues killed in action or taken prisoner.

When alcohol flowed the stress of operational life was eased and pubs became the centre for pranks and shows of exuberance. Off duty airmen rarely talked about their missions but regularly boasted of their daring deeds in car and motorbike races along local roads, or in drinking contests in the four ale bar.

Not all servicemen found solace in pubs and not all had recourse to a platonic female friend. For many the solitude of the countryside surrounding their place of work brought its own peace and comfort. Peggy Newton was living with her family in a farmhouse on the flightpath for RAF Methwold. She regularly lay awake in bed counting the planes out and back. In 1943 on a raid much like any other, twelve Ventura bombers took off for the Dutch coast to carry out a diversionary raid on the power stations at Amsterdam. One aircraft returned to base early with technical problems but the remaining eleven flew into an ambush. Ten aircraft from the squadron were shot down, one which was seriously damaged returned to its Norfolk airfield. Soon after Peggy was walking near the 'drome. A man was walking towards her. She immediately recognised him as a crew member from one of the planes that had returned. He did not see her. She remembers that his eyes were glazed. He stared straight ahead of him, walking with automatic movements, unaware of anything or anyone around him. She could only guess at the grief and guilt he carried at having returned when so many of his colleagues had not.

When the worst did happen it was the job of a surviving member of the crew or senior officer to write to members of the dead man's family. As adjutant, Edgar Reeks had his share of letters to write and funeral arrangements to make. He recalls one incident."There was an explosion one night. A plane had crashed, killing all six members of the crew. There was a full military funeral but later there came a letter from one of the parents accusing the RAF of robbing their son of his wrist watch, which had been a twenty-first birthday present that he would never have parted with. My job was to tell his parents that all we could find of the six bodies was two pounds of flesh which had to be divided between the six coffins."

H Grantham Hill remembers funerals in country churchyards where full honours were bestowed with a detachment on escort duty and a volley of shots over the coffins. As the boxes were lowered only a few of those present would know the truth that they contained stones, since there had been no bodies found to bury.

On American airfields there was a small mortuary on site but as soon as possible all mortal remains were despatched to a unit called G23, based in Cambridge. A funeral detail, led by the base chaplain, would then be sent to the United States cemetery just outside the city at Madingley. Louis Pennow remembers one particular occasion when funerals took place amid pouring rain. "It was depressing enough to go to a funeral, but to go to one in the wet is particularly nasty," he says. That day twenty-two men were being committed to the ground, ground which was a veritable sea of mud covered with boards which led to each open grave. "So many people were being buried, week after week," he recalls, "that the ground never settled. The mud was created by the rain and the funeral parties. At the end of the day our trousers were packed with mud to our knees where it had splattered back from the planks."

By the end of war in that one cemetery there would be more than five thousand graves.

13 Local Effects

Today there would probably be outcry if word got round that a new town was to be developed in the midst of peaceful countryside and a population in excess of a thousand people be imported to live there. Petitions would be drawn up complaining at the noise nuisance and the unsuitability of local roads and facilities to cope with the demands such a development would create. There would be public inquiries and questions in Parliament. And yet all those years ago, when plans for an airfield were unveiled the effect to the local area was as profound as if a new town was being built – more so because most towns did not deal in death. Yet in 1939 and during the war years, there was very little consultation with local people who would find themselves with a new neighbour and no opportunity to make known their feelings about the effect it would have on their quality of life. Inevitably there was antagonism and bitterness, particularly where compulsory purchase orders had been enforced or rights of way had been lost.

At one end of the inconvenience scale were householders whose gardens suffered because of the high spirits of servicemen. Tales abound of flowers going missing from front gardens, babies being woken and livestock frightened as cars and motorbikes competed to be first back to the station guard house. At the other were landowners who initially felt it was unnecessary to acquire valuable agricultural land for the purpose of building an airfield. The price paid for farm land acquired thus was less competitive than during the early 30s expansion period and some farmers felt that they had lost out in financial terms. But as war progressed events moved quickly and feelings changed with them. In 1939 the idea of England being in any form of danger and in need of airborne protection seemed remote and our bombers could fly from airfields in France. The evacuation of the British Expeditionary Force from the beaches of Flanders and increasing attacks over the English mainland quickly challenged unrealistic notions of Allied supremacy and the fact that England was, somehow, beyond the enemy's striking distance. As events unfolded the earlier feelings of resentment at having land taken for airfield construction changed to one of satisfaction at helping to play a part; especially when city after city suffered under the wings of Luftwaffe planes bringing war to the homes of many who thought they were safe and well behind the battlefront.

There were advantages too in having an airfield in the neighbourhood. Many a local taxi firm benefitted from journeys transporting servicemen between their camp and the nearest railway station. Local traders and suppliers found themselves with a sudden increase of several hundred customers. The local economy began on an upward trend.

Suffolk farmer John Wortley found the patronage of his local airfield a godsend during the troubled days of wartime. Where he had once supplied 150 gallons of milk to the nearby town of Thetford, he was suddenly called upon to provide 500 gallons for Honington. The sergeant's mess there was a customer for his fresh milk and normally took a daily order of 45 pints. He recalls one day when the squadron returned from a raid on Hamburg having lost eleven aircraft. The order then was reduced to a mere five pints. There was also contract work available to local farmers, who were issued with special passes to gain access onto the airfield to cut grass or help with general maintenance.

At the rural railway stations where George Coleman worked there was evidence of airfields in the vicinity which went beyond the sudden increase in passengers with 48 hour passes dressed in Air Force blue. When the local airfield was staffed by Americans there was an increase in the number of female passengers using the station's facilities, most of them originating from the London area. There were always American servicemen waiting at the ticket barrier to escort the ladies and accompany them back in time to catch the last train home.

Increased rail traffic at evenings and weekends as servicemen travelled on leave or to local cities meant extra work and overtime for many railwaymen. In the early days of airfield construction railway goods yards were often hives of activity as building materials arrived by rail. George recalls trains of nearly one hundred trucks arriving laden with Sommerfeld Matting and pierced plank steel tracks for use as runways.

"Men from the airfield spent a few weeks at the station unloading the landing mats. Each day, seven days a week, a vehicle would bring hot meals. There was always plenty and the men would make sure there was some for us railwaymen." says George.

That generosity of servicemen towards local civilians was the typical experience of many countryfolk with their new neighbours. When those neighbours were from across the Atlantic the benevolence was greater. "I remember I pumped up a bicycle tyre for one private and he gave me a half-crown tip (12½p) which was worth a great deal then," recalls the railwayman. "When train loads of men arrived at the station there was always food travelling with them. When the carriages were unloaded they would always give a tin of biscuits or fruit to us."

The arrival of an airfield in the locality was always of special interest to boys in the parish. In a later chapter the close bond that developed between an English schoolboy and the American servicemen on a nearby airfield is described. Such a bond and fascination was not the exclusive domain of East Anglians and Americans. Neil Graham was ten years old when war started and before he reached his twelfth birthday his home village in Yorkshire was all but surrounded by airfields.

"In 1940 we moved from Thirsk to Topcliffe. RAF Dishforth had been built before the war and RAF Topcliffe was nearing completion. Soon landing grounds at Dalton and Skipton on Swale would be built and Topcliffe village was ringed by four airfields," he remembers.

"At first we saw overhead Whitley twin engined bombers which always seemed to fly 'nose down'. They were slow and it is difficult now to think that we expected brave young men to fly them by night, carrying bombs, reach, find and bomb a set target, avoid anti-aircraft fire and night fighters, get the navigation right, have the luck to keep the engines going, and find their way home. Not many had the luck to do it for long.

"The Royal Canadian Air Force then took over at Topcliffe. Here was another modern marvel – real live young men from the other side of the Atlantic! Some Sunday evenings a score or more of them would be at chapel. We got to know some of them very well. We sang hymns round the stove and the piano after evening service.

"A few men were aircrew. We never seemed to be given the time to get to know them. Losses in men and machines were heavy."

He remembers the planes taking off, heavy with fuel and bombs. He was not always awake when they returned home and it was often his mother who would tell him at the breakfast table that not many aircraft had returned.

Neil Graham remembers to this day his heartfelt response, which was always the same sincere and fervent hope that "Perhaps they landed down south".

* * * *

Among the negative effects of an airfield in the neighbourhood was the threat to public safety caused not only by enemy raids and attacks, but by routine operations. At one of the stations near Neil Graham's Yorkshire home a Halifax bomber crashed just half a mile short of the runway at Dishforth when it ran into a small fog of steam from a locomotive on a nearby rail line. The plane was on a training flight from Croft and of the six crew members, five were killed and one injured.

At Little Walden in Essex a plane crashed soon after take off and a civilian perished in her desperate efforts to rescue the crew, earning her a posthumous award from the American Air Force. In January 1945 an American Liberator bomber heading for Horsham St Faith after a bombing mission on Halle crashed into the garden of a council house in Norwich. Two children playing outside were killed and the entire crew save one member were lost. One child playing at the spot seconds before the crash survived. The crewman who lived never flew again.

Over the same city 83 mines were accidentally discharged by an allied aircraft and elsewhere in Norfolk at East Wretham a Wellington bomber crash-landed in flames on a local road, the bomb load still on board detonating soon after. Across the county border in Suffolk, a bomber struggling to get height at Framlingham fell from the sky and crashed on a row of houses. The entire crew perished and despite every house in the terrace sustaining damage there were few civilian casualties.

And even when it was safe it was often inconvenient to have an airfield nearby. There was the American crew of a B24 who on take offs and landings always flew perilously close to a local windmill, hoping their propwash would cause the sails to turn. At Topcliffe Neil Graham remembers the Black Bull public house, a favourite haunt with aircrew, regularly being 'buzzed' by Wellington bombers. It became something of a tradition which only ended when the local vicar wrote to the Commanding Officer expressing his fears for the safety of the church tower.

At Pocklington there was the noise nuisance which led local villagers to make representations to the station about their interrupted sleep patterns. Fight the war, bomb the enemy, bring him to his knees, but please do it quietly! When lights were on at those stations where FIDO was installed villages within a mile of the runways benefitted from a sudden burst of street lighting. "You could see to read a newspaper and the birds thought it was daytime," remembers a resident who lived near Carnaby in Yorkshire.

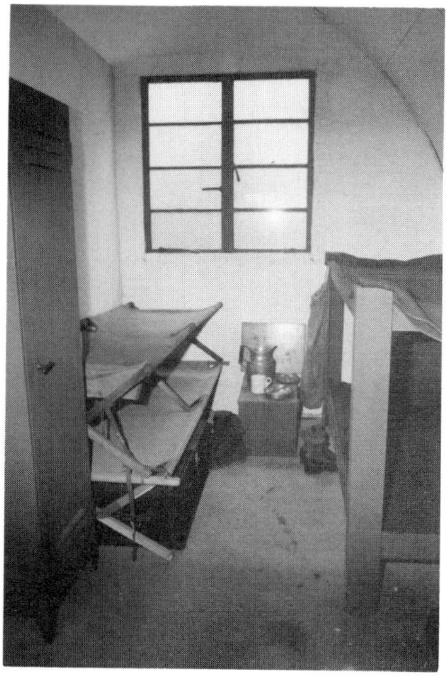

Far left *Reconstruction of an American Eighth Air Force airmen's billet at Thorpe Abbotts.* (Author)

Left *A flight maintenance wall board at an American airfield giving up to date details of aircraft being serviced or repaired. A reconstruction at the 100th Bomb Group Memorial Museum, Thorpe Abbotts.* (Author)

m Hurry, resplendent in new clothes with members of the groundcrew he helped at Thorpe Abbotts. (Sam Hurry)

Nearly fifty years on, the pull of Thorpe Abbotts on Sam Hurry is still strong. He is a leading member of a team who have restored the former control tower and nearby buildings into a permanent museum and memorial to members of the 100th Bomb Group who flew from this South Norfolk airfield. (Author)

14 Over Here...

Perhaps the most profound effect of an airfield on the neighbouring communities, was the sudden influx of strangers as servicemen and their families moved into the area. The areas where stations developed were often in the heart of the countryside where villages were isolated, public transport poor and car ownership limited to the wealthy. Rural communities then were settled places where people still moved house during the Michaelmas hirings, if at all, and often then only to an adjoining parish. Registers and log books from village schools during the first forty years of the century record the same family names passing through classrooms, generation after generation. In some country areas the arrival of children from cities under evacuation schemes was shock enough to many a countryman's constitution and shyness, but when an airfield arrived the shock was intense. Not only was there a sudden increase in population caused by the newcomers but many of them came from places thousands of miles away.

When war was declared Britain's empire and colonies rallied to the flag and mustered units of men to serve in all branches of the military. But it was in the Air Force where foreign nationals found themselves living in the midst of rural communities in which any newcomer was viewed suspiciously, that the differences in culture and life style most apparent.

At Driffield, Yorkshire, a squadron of Halifax bombers was based crewed entirely by Australians. Inevitably one of the aircraft was christened Waltzing Matilda. Britain's allies offered help and in the same county at Elvington the French air force shipped in Algerian wine to accompany meals which, although cooked by French kitchen staff, consisted of English cuisine. Canadians were based at many stations, including Coleby Grange in Lincolnshire where one pilot lived in a country house, playing the part of an English country gentleman and riding his own horse around the village. In Suffolk life was even more cosmopolitan with squadrons of Polish and Czech airmen and aircrew. The registers at the training college at Cranwell give the best indication of the range of foreign nationals training there to help further the Allied cause. During the wartime period records show the places of origin of the young pilots and navigators, including among them every Commonwealth country, plus Ceylon, Belgium, France, Czechoslovakia and Turkey.

But the biggest impression of all the overseas visitors to English airfields at that time must surely have been created by the Americans. "Over sexed, over paid and over here," was a popular adage describing the situation for many an Englishman enduring rations, meagre pay, wartime conditions and a wife or girlfriend far away. Certainly the lavish lifestyle of the Americans endeared itself to many. In austerity Britain where everything was rationed or only available at extortionate prices, the ability of American servicemen to acquire seemingly unlimited stocks of candy bars and nylon stockings became legendary. The American economy contrasted with that of an island economy already stretched by war. Supplies and commodities were brought in by sea and prone to attack by marauding U boats. Goods made at home had to compete with the need to maintain manufacturing at constantly high levels to feed the country's insatiable war machine. There was scarcely enough of anything to go round; quality of goods dropped and food substitutes abounded. But none of this affected the Americans who shipped supplies in from the States. In many cases local people benefitted from their generosity.

Railwayman George Coleman recalls a lorry arriving in his village from the nearby airfield where Americans were based. It stopped at pre-arranged points around the place to collect children for a trip to the airfield where they would take part in a Christmas party as guests of the American airmen. At other times their lavish lifestyle contrasted with that of their host country. While the message to Britons was 'Make do and Mend' the visitors had enough of everything to be able to discard it long before it had outlived its usefulness. George Coleman speaks of the American lorry which made a regular trip to the local dump laden with goods which were excess to requirements on the airfield.

"There were all kinds of things on board," he remembers. "There were many empty boxes but there were some good boots, some new overalls and brooms and mats. It seemed to me that when the men who cleared up in the billets reached the door, the mat and broom were thrown away with the dust and rubbish! And if boots had been laid out on the mat, they went as well."

There was a darker side to the contents of this particular vehicle as he explains: "Needless to say there were always folk waiting for the lorry at the tip. I remember well some German prisoners of war working on a field nearby and going over and discovering parcels of cigarettes, cigars and chocolates. They were all unopened. They all had labels on, addressed to servicemen and bringing love and greetings from families, wives and girlfriends back home. We were told they had been sent to airmen who had not returned from their missions."

The Americans preferred to fly in their own produce and John Wortley remembers that his contract to supply Honington was not required when the RAF moved out and USAF moved in. "Things changed considerably from a tradesman's point of view. Our meat, milk and potatoes were not good enough. All must come from home – in cans."

He experienced the squirrel mentality of the American Air Force as it imported its needs and built up great stock piles. Airfield buildings were no longer able to accommodate the vast stores and supplies accumulated and extra storage areas were built. Even this did not appear to meet their needs and the Suffolk farmer recalls his surprise when he was a member of a group clearing up on the station after war had ended and the Americans had pulled out. He excavated one of the pits which had been dug when the airfield was first constructed and found it full with parts of aircraft engines, armaments, timber and oil cans. From three similar pits a veritable hardware store was uncovered: an unused electric bandsaw, forty new tarpaulins, about 100 wooden planks twenty feet long, pieces of plywood and sundry electric motors and tool chests. Also discovered that day was a pit where the bodies of 30 cats and dogs had rotted away.

As well as their supplies, the Americans imported their own culture. In village halls jitterbugging took the place of sedate waltzes and cheek to cheeks. For station concerts they flew Glenn Miller in from the States.

The American sense of identity was strong and on occasions the two cultures clashed, as occurred on the airfield where Edgar Reeks had become part of a small British administrative presence. The station was in use by the American Air Force but remained under the control of a Royal Air Force group captain. Day to day domestic arrangements were left to the visitors, who fixed eating times in the mess between 1800 and 1900 hours each day. A 24 hour service was available from the sergeant's mess. One memorable day the group captain came in for dinner five minutes late to be greeted by the American at the servery calling him 'buddy' and referring him to the 24 hour service at the sergeant's mess. The disgruntled group captain immediately called over an American officer to put the cook in his place and explain his rank. The American officer looked at the four rings on the British officers uniform in amazement and before the group captain had a chance to explain the snub which had just occurred, the officer enquired of him: "Wow. What section do you work for, buddy?"

Louis Pennow arrived in the United Kingdom to operate a headquarters based weather station. As part of his training he served on airfields at High Wycombe and Brampton, moving on to Knettishall where he was part of a detachment of men reporting flying conditions to a central office. Like H Grantham Hill, whose work as a met officer on RAF stations is described earlier, Louis helped in the briefing of crews before a mission, and debriefing on their return to check the accuracy of the forecasts.

He lived in a Nissen hut accommodating 24 men arrayed in two tiered bunks. The main feature of the hut was the stove. Often the tortoise stoves were replaced by a 'potbellied' burner brought from home, usually with a coffee pot boiling away on top. There was never any need to worry about coal supplies to keep it stoked up. The Americans soon discovered that compressed paper packaging used around the bombs had excellent burning qualities. When the week's ration of one hundredweight of coal had lasted the customary two days the stove was filled with paper packaging and with skilled use of the draught slider, the blaze would soon be roaring away. The entire stove glowed cherry red and the sheer heat would prevent anyone standing too close. The only way to keep the blaze stoked up was to wait patiently for it to cool.

Leisure time was spent in the day room where table tennis, poker, chequers and a piano were available. There were three clubs available: one for officers which contained a bar; one for sergeants with a dart board; and an American Red Cross club for all ranks which served coffee and doughnuts. Louis Pennow's weekly pay was in pounds sterling and was often spent at the station post exchange (always known as the PX). There Mars bars, Dentyne chewing gum, 'shoestrings', dusters and toothbrushes were on sale. The ration system allowed each airman seven packages of cigarettes but there was never any choice offered: whatever happened to be in stock was doled out, regardless of personal tastes or preferences. He remembers the run up to D Day when the UK was full of GIs waiting to step into Europe. The cigarette stocks at that time consisted of brands never heard of before and the quality of tobacco he likened to sweepings from the warehouse floor. Even the land of plenty, it seemed, had trouble meeting demand when needs were great.

The British Church Army sent around a converted van selling cups of tea and cake at a penny each. It was at this time the American met officer developed his taste for English chocolate. Although Hershey bars were imported to airfields from the States and always proved popular with local children, Louis Pennow keenly waited for those occasions when the base PX received a consignment of Frys Chocolate Cream bars.

Close ties developed with local families and many American servicemen were accepted with almost the same ease as relatives, as though they had a birth right to be there. Children whose fathers were away overseas or posted miles away from home valued the men's presence in their homes and villages. When a three day pass was issued and hospitality was offered in a local home, Louis would visit the Mess Hall and collect a tin of Spam or pineapples in syrup: some token to help augment the host's food ration. Yet for all those local links, there could be no substitute for the home country and for many servicemen a visit to a Glenn Miller concert was next to a letter from home.

Louis chose not to think too much abut the family he had left behind. He sent off letters regularly enough but was aware that obvious home sickness could affect his task. Too much time and effort thinking of home by too many people could easily affect morale and unit performance. "My family were the guys I was with," he states. "My loyalty was to them. Most guys were well adjusted to life far away from home. Before arriving in the UK we'd been in service for a year or two. We were busy doing a job, we made the most of it. It was hardest of all for those who were newly married with a baby back home."

Close friendships developed within each unit, the more intense for being forged far from home in conditions of war. Home leaves were not always possible, even if a parent was ill or a wife about to give birth to her child. At best the system would allow for small changes to duty rotas enabling servicemen with domestic worries to stand down for a few days so his anxieties could not spill over and affect his task. Applications for leave were made but they invariably came back with a TS ticket attached – an abbreviation standing for Tough Shit, or, as the English would have had it; "regrettably the application cannot be granted."

* * * *

It transpired that one of the largest invasions of the war was also one of the most peaceful: the arrival in the United Kingdom of the American Air Force with their brief to assist British forces in the struggle against Axis powers. From 1942 they began to arrive, moving into new airfields, taking over the lease of others as English squadrons moved out. At some locations they constructed their own airfields from scratch using their own engineers and equipment and a large force of British labourers to get the work completed on schedule.

In Norfolk alone 100,000 acres of agricultural land were taken over by the United States Army Air Force and in some parts of East Anglia the number of Americans outnumbered locals. Many had left behind well appointed training schools and found it hard adjusting to the primitive facilities afforded by operational airfields. As with English airmen they spilled into the towns on off duty periods and came into close contact with the local population.

Remembering the advice in the booklet they were issued with on arrival here, they tried to be friendly but not obtrusive. They were told that they were better paid than their hosts and shouldn't rub the fact in. "Play fair with him, he can be a pal in time of need," advised the American booklet about their British hosts. They were warned not to make fun of local speech or accents. "You sound just as funny to them but they are too polite to show it," said the text.

Some crews arriving at airfields were fresh from training school and lacked any combat experience. The early strategy adopted by the American Air Force was daylight bombing of enemy territory without any fighter escort. Although they were well equipped with aircraft, a combination of pilot error and inexperience and the tactics themselves resulted in huge losses. In the last three years of war some 46,000 men and 6,000 aircraft were lost by the Eighth Air Force in England.

Such were losses that to complete a tour of duty became a real cause for celebration on the airfield. The 100th Bomb Group had their own tradition for those who managed to cheat death and after serving a stipulated number of missions they were enrolled into the Luckye Bastardes Club. It was a brotherhood with its own exclusive membership parchment awarded to each member for "transporteing far and into ye German Reich full manye a tonne of petardes, mines, balles of fire and other grievous burdens for ye discomforting of ye manne Hitler and alle his eville companye."

Examples of heroism vie with tales of incredible adventures experienced by aircrew. The story is told in Suffolk of a crew who became separated from their main force. They decided to continue the mission alone but while trying to locate the target were attacked by German night fighters. Jettisoning the bomb load they abandoned the mission and turned for home, eventually spotting what they took to be the coastline of England. As they flew along the line of cliffs and shore, it soon became apparent this was not home territory. Consultation and calculation revealed that somehow they had completely managed to avoid flying over mainland Britain and had arrived in Ireland. Despite turning due east they were unable to reach their East Anglian base, a shortage of fuel forcing them down in Scotland. Small wonder on their eventual return south of the border an RAF navigator lost no time in suggesting they attached a reel of cotton to the wings of their B17 to help them find their way home on future missions!

On October 8 1943 planes from 100th Bomb Group set off on a mission to Bremen during which seven planes and 72 personnel would be lost. One aircraft that made it back was called Just A Snappin. Caught by heavy flak, the plane and its crew had little option but to break formation and return home. In doing so they were subjected to enemy fighter attacks causing further extensive damage to the airframe and injuries to four crew members.

The personal report filed by the lead navigator gives some idea of their plight. "We had survived everything that the Germans could send up at us but there was still one thing to consider – gravity. Crossing the coast had cost plenty and we were now at 3990 feet and sinking rapidly. . . . Ditching seemed the next answer . . . England seemed so far away. The ship was listing in such an attitude that our . . . compasses stuck on the side. We hit England. By now our gas problem was serious.

"The very first airport sighted at Ludham seemed large enough . . . so we prepared for a crash landing. On the entire field there were two big trees together and the plane now completely out of control veered towards them. The largest one hit while the plane was going about 50 mph. Luckily it passed between the second engine and the pilot's compartment so we were swung round instead of being jerked to a halt."

Medical aid was two hours away and eventually two ambulances arrived to carry the injured to the Norfolk and Norwich Hospital. Many other planes on the mission were so badly damaged they never returned to their base and it was a long time before the Group commenced normal operations again. In human terms the crews needed time to recover from the shock of losing so many friends and colleagues. On a purely practical level, new aircraft to replace those lost in the raid were slow in arriving.

When life in the air was hard and stressful opportunities to rest, unwind and relax were important. Christmas for the crews of 100 Bomb Group in the wake of the disastrous Bremen raid saw a festive meal provided consisting of turkey with cranberry sauce, snowflake potatoes, creamed peas, Waldorf salad and sweet corn, followed by apple, raisin and mincemeat pie. The belief was that if you worked hard, you played hard too. Great celebrations took place on September 30 when the 200th mission was flown. Athletics races were held around the perimeter track with twelve prizes including 'half a case of liquor'. Aircraft were wheeled away from hardstandings and hangars cleared. Fairground rides were erected on the runway apron and trestle tables were set up heavy with beer barrels. An old fashioned barbeque followed, with dancing in the hangars to no fewer than four bands.

Often local people were invited to the celebrations, an occurrence which rarely happened at RAF stations. Discipline and security was slacker than that imposed on British airfields as war progressed. In the air some people felt that the standards governing the RAF were higher than their American allies. Incidents are recorded of crews forgetting their recognition signals resulting in anti-aircraft gunners opening fire on them as they returned from raids. Yet, on the ground they were ordinary people, nursing the same hopes and fears as those in the cottages and farms around them. They were outgoing and friendly, their extrovert and often flamboyant life styles taking their reticent country neighbours by storm. After initial suspicion, fears were calmed and locals were soon won over by the newcomers' generosity. For children whose fathers were away fighting some of the Americans adopted a fatherly role, and for some women the ties were close enough for 50,000 to marry American servicemen posted here.

Louis Pennow was among them, courting a girl from Cambridge and eventually marrying here. They returned to the States but after a few years crossed the Atlantic again, this time to make a permanent home in Britain enabling Pennow to continue his working life on an English airfield used by the United States Air Force in Europe.

15 The Boys Next Door

On an evening in 1942 a schoolboy lay in bed fighting off sleep to listen to the muffled voice of his father downstairs giving news that there was to be an airfield built a mile or so from the village where they lived. Those snatches of conversation marked the beginning of an association between the schoolboy and the airfield which would last long after war had ended. It heralded a time in the boys life which would see him unofficially 'adopted' by American servicemen who welcomed him sharing their routines of work and leisure, providing him with a variety of experiences and insights deeper and broader than any education the tiny village school could provide.

So, from April 1943 when the first Americans moved into the parish of Thorpe Abbotts, the school attendance records of young Sam Hurry became decidedly erratic, and for some periods of time, non existent. Most days during term time and every day during holidays Sam and his friend Billy Taylor would be somewhere near the airfield. More often than not they would be on it, working with servicemen. Although Sam's home was a bike ride away, Billy's cottage home became neighbour to a dispersal and fuel loading position. A home so close to the airfield's flying operations was just the base the two eager and inquisitive boys needed.

For Sam Hurry the fascination for the place started early on in his life. He gathered acorns here to feed the local pigs, knew each and every tree and which part of the pond was the haunt of the biggest fish. Quite suddenly new people arrived and began to change the look of his familiar countryside: builders and labourers billeted in local homes, trucks and lorries negotiating narrow cart tracks and lanes. Then the Americans arrived.

"I can remember them coming quite clearly," he says. "Billy and I were near the control tower. There were already a few Americans there. They were a skeleton staff, getting the place ready. Suddenly the sky seemed to be filled with great aeroplanes. We had never seen planes as big as these with their four engines. We stood and watched the huge bombers come towards us then circle before landing. It was some time before we dared leave the spectacle and go back home."

He wonders what the first reaction of the visiting airmen was to the two local boys. "We must have looked like strays," he recollects. "We were two forlorn looking, half starved characters scruffily dressed in torn clothing. They asked us if we had eaten that day and the answer we gave was 'no'. We were then lifted into the mess wagon and taken to the mess hall near the Red Cross building. Boy, was I scared. I had never seen so many Americans in one place, nor so much food before.

"We waited in line with mess can, plate and eating irons. When our turn came meat loaf, potatoes, carrots, peas and gravy were put on the plate. Then, to my horror, slap onto the same plate came peaches and cream. I was astounded at being served all this on the same plate but we soon overcame our shock. The food seemed as if it loved mixing together and we travelled many times to the mess hall."

The Americans' arrival heralded an upturn in the local economy and the boy's mothers were quick to realise they could meet the airmen's need for clean clothing by operating laundry businesses. The laundry run by Sam's mother and sister in law was typical of a number in the area set up in local homes. It boasted a clientele of between 50 and 60 and turn round time for clothing was two days. Sam did the fetching and delivering, walking from his home to the airfield and around the billets there pushing a handcart.

On days when the laundry was running the whole house would be full with steam and on wet days clothes hung inside from hastily strung up linen lines or were draped across pieces of furniture, creating a strangely draped obstacle course through the familiar rooms. In Billy Taylor's home, drying presented fewer problems. Being so close to a dispersal point it only needed a word to groundcrew to test the engines of any plane they were working on and the draught caused would be enough to thoroughly dry a complete line of washing in ten minutes. When laundering was complete it was delivered back to the airfield, and new items for washing were collected. Many times a laundry bag or pillow case full of clothes would be turned out on the table of the home laundry and there at the bottom would be a tin of fruit or ham as a gift from their American neighbours.

The close proximity of Billy's house and regular visits of the laundry handcart led to a growing friendship between the boys and the airmen serving at the dispersal. They allowed him to assist in their duties and Sam remembers helping wash down dispersals when aircraft had gone off on mission. The hardstanding was covered with 100 octane aviation spirit and the youngster would be given a lighted rag to throw on it. Immediately the dispersal burst into flame and as it did so all the oil and grease would be burned off, leaving behind a clean working surface ready for the aircraft to return.

A highlight of his days was to be invited by groundcrew to sit in the cockpits of the huge B17 aircraft while engines were being tested. Instructions would be shouted up to the schoolboys who had already been instructed in the procedure of throwing switches to fire the engines into life. On those occasions aircrew remained unaware that the boys had been inside the aircraft, but there were times when even the ground crew did not know that Sam and Billy had taken to the flight deck and to the accompaniment of sundry throaty roars seated themselves at the controls and proceeded to win the air war single-handed.

A further pleasure was to sit on the wings of a bomber being serviced at the dispersal. It was an ideal place to sunbathe but sessions ended when a ground crewman told them to remove their studded boots in order that they should not damage the outer skin of the plane. Alas, Sam's socks were full of holes and pride would not permit him to remove his shoes so those hours spreadeagled on the wings soaking up summer sunshine came to an abrupt end.

The Americans had a sense of fun and frequently used it at the expense of local folk. It was not unknown for them to wait until a local resident was passing by on a bicycle and then choose to run an aircraft engine at high revs to see if the wind created was sufficiently strong to unseat the cyclist. But Billy's mother was never the victim of such pranks, indeed not even the might of the American Eighth Air

Force could stop her from cycling across the airfield to a whist drive in a village on the far side.

"She was a keen whist player and most weeks during the war she would get out her bicycle and go over the airfield to these get-togethers. Quite a normal thing to do, until it was realised that the route she took was directly across the main runway. She had no fear of aeroplanes. They would be taking off while she happily rode behind or in front of them. She could often be seen riding her bike on the perimeter track with an aircraft behind her," he remembers.

When the Provost Marshal, chief of the airfield's military police, intercepted the venerable High Emma cycle and its rider on one of their weekly crossings of the runway he was left in no doubt exactly who had been there first!

On autumn evenings Sam would help lay the flare path for incoming planes. At that time containers resembling watering cans were used for the purpose. His job was to sit on the tailgate of a truck and light the wick protruding from the spout of each can. Once burning he would pass the can to a serviceman walking beside the truck who would place it in its correct position. Each evening on flare path duty Sam would start out clean and polished and finish covered in black smuts and reeking of oil. The problem was particularly acute on foggy nights when the damp atmosphere trapped the smoke from the burning flares around him. When the flarepath was laid he would be taken to the control tower or ambulance shed to wash before going to the mess for a meal. By this time it was often late and to avoid a cycle ride home through the blackout Sam would stay with Billy, the better to be up bright and early for another days service on the airfield.

That service was recognised and appreciated by the Americans and as a thankyou gesture he was treated to a day 'on the town'. Groundcrew at the dispersal where he helped out contributed money to kit the schoolboy out in new clothes. Since they had no spare clothing coupons they paid expensively for items at the local Co-operative store. It was as though Christmas had come early that year when they presented him with three pairs of shoes, socks, trousers, a lumber style jacket, a tie of Scottish plaid, cap, shirt and underwear. And the treats did not end there. That weekend saw him ready to undertake his first ever trip outside his home county. As a guest of the American groundcrew he was taken by train to London to tour the zoo and spend an evening at a theatre. And of all the things he saw that day, in theatres, animal houses and the bomb sites of blitzed London, it was his first sight of a flush toilet that impressed him more than anything.

When not working he would lay in grass at the end of a runway and watch the B17s as they lumbered into the sky. "It gave us a sense of belonging to the place, and a possible ticket to the next," he now says wryly. Other days were spent furtively sitting on bombs in the bomb dump, smoking cigarettes and chalking messages and inscriptions on the bomb cases.

At times he himself admits he was more a hindrance than help. He had been given a cigarette lighter by a member of the ground staff who recognised that at ten years old Sam was already a seasoned smoker. He was now the proud owner of a Zippo type lighter, and told to fill it with petrol from the 50 gallon tank placed near the hardstanding. This task accomplished he sat down beside the tank and began the procedure of lighting up. Within seconds the tank was alight. Thanks to speedy reactions from the ground staff it was quickly extinguished, only to burst into flame later and burn down a tent.

The 100th Bomb Group based at Thorpe Abbotts suffered heavy losses and each was felt keenly by young Sam Hurry. Although he did not know the aircrew he and Billy Taylor were regularly by the runways at seven in the morning counting out the planes as they left on their daylight bombing raids. And they would be there when they returned, counting, waiting, hoping . . .

Often such hopes were in vain. He remembers one day waiting for aircraft to return from a raid on Munster. He and his friend stood on the perimeter track at the time they expected the force to return. Nothing happened but they continued to wait. From where they stood they could see that all was not well at the watch tower, could sense an atmosphere around them heavy with foreboding. They could see the destination board in the watch tower and the names of thirteen planes that had left Thorpe Abbotts. As they waited there came the sound of aircraft engines but it was not the heavy roar of skies echoing with returning planes. Only one plane appeared from behind clouds and landed at its home base.

Dispersals which that morning had been full now stood empty. Groundcrew remained beside jeeps and trucks, waiting for their particular ship to return. Personnel carriers still circled the perimeter waiting to collect crews from aircraft they still hoped would come home. Americans who had been characteristically happy go lucky and devil-may-care in their daily routines no longer smiled or laughed easily. There was no longer any spark of life, everyone was devastated. It was a full week before the dispersals would be occupied by replacement aircraft, a week before the mess buzzed with the enthusiasm of new crews. Things were never quite the same after that day.

Occasionally moments of danger and alarm were tempered by lighter moments. He remembers a mid air collision between two aircraft directly over the airfield which caused the wreckage of one plane to fall on the bomb dump in the south west corner. An unofficial evacuation ensued with servicemen commandeering bicycles and heading at speed away from the crash site. Some personnel used the incident as the excuse for some unofficial leave, and during the confusion managed to leave the airfield. The story was told of a GI spotted locally pushing a bicycle at the double along the lanes. When asked why he wasn't riding it he replied that he hadn't had the time to get on it! Fortunately for all personnel, the wreckage when it fell missed the high explosive store, landing instead on the incendiary dump. By nightfall the fires were under control and a near disaster had been narrowly averted.

On another occasion a Liberator bomber overshot the runway, landing in the local pond where Sam and Billy had spent pre-war days fishing. It would be another fifteen years before the contamination caused by aircraft oil and engine fluids had cleared and the pond would be fit for fishing again. Another day a crew were joyriding in their plane but in lifting its nose to clear the runway failed to gain the necessary height, clipping a tree and crashlanding beside a local farm. The plane, ironically named Hang The Expense, was written off and a bull in a farm shed demolished by the impact was killed. There was no loss to human life.

But it wasn't always like that. Sam watched many aircraft landing in difficulties, including one that came in with only two of its four engines running. One engine then proceeded to fail before it landed and the other cut out immediately the plane came into contact with the ground. The aircraft slewed to a halt where it was met by the base ambulance. The injured were brought out wrapped in blankets and when he walked past the camp hospital he could hear their screams.

When the Americans pulled out in December 1945 Sam Hurry was nearly twelve years old. He had missed much of his formal schooling yet living on the doorstep of an operational airfield and developing close and lasting friendships

Right *A mural of a Stirling bomber was removed from North Creake airfield and has been relocated at the RAF Museum, Hendon. Here Mrs Dorothy Allen, widow of Flight Sergeant Ted Allen who painted it, sees her husband's painting for the first time at the unveiling in 1984.*
(Eastern Counties Newspapers)

Left *Runways, hangar and nissen huts remained at North Creake after the war, to be joined by new buildings when an agricultural firm took over the site.* (Brian Beaumont)

Below *The perimeter track of RAF North Creake runs across the left hand corner of this view. The scorched square beside it was the site of Brian Beaumont's home demolished when the airfield construction began. The semi-detached houses in the centre were built to replace them after the war.* (Brian Beaumont)

with those stationed there had given a broader education than any classroom could offer.

He knew from bitter experience the taste and side effects of salt tablets, illicitly purloined from a pack of K rations, which had left him drinking water from the dyke so great was his thirst. He knew the look of men facing death each day and came to know how it was to mourn for friends who never returned. There was the fun of sunbathing on aircraft wings, then the horror of seeing those same planes landing, riddled with shot and damage, belly flopping across the airfield, paying no heed to runways.

Thorpe Abbotts where these incidents occurred was part of a huge complex organisation that was the American Eighth Air Force. It was a place of joy and sorrow. It was a place that etched itself on the heart of one English school boy as surely as had he been a young conscript into the American Air Force.

16 Egmere Airfield

When war broke out the Norfolk towns of Wells and Fakenham were joined, as now, by two main routes. Known by the locals as the 'wet' and 'dry' roads, one passed through villages where travellers could find hospitality in public houses, and the other through largely unspoiled countryside where there were none. Indeed, there was very little in the way of settlements along the dry road and one of the largest was at a small country cross-roads where a few houses had been built one side of the junction. These houses, together with a few scattered farms and their attendant tied cottages, comprised the hamlet of Egmere. There was no shop, no church, no school. Across the fields in each direction were the villages of North Creake and Walsingham, to the north the small coastal port of Wells and to the south the bustling market town of Fakenham.

The residents of Egmere worked mainly on the land. It was a solitary way of life governed and checked by weather and passing seasons. In the summer the dry road was busy with cars and charabancs heading to the beaches at Wells but with war the entire coastline and a corridor twenty miles inland had been declared a no-go area where regional civil defence controllers had power to check all people entering the area. Such was the fear of invasion from enemy occupied Europe, government restrictions meant it was now possible for men to work in the fields and not see another person until the days work was done.

This did not unduly worry young Brian Beaumont who lived with his parents in a farm cottage. He had all the friends he needed a short bus ride away at the village school in Walsingham and in the belt of trees near his home there was a regiment of soldiers from Newfoundland. They manned two large guns pointing across the fields out to the North Sea. The guns were tested regularly and with one sited only a hundred yards from the house messages were relayed to the occupants to open all doors and windows. No chimney sweeps ever needed to call at the cottages while the Canadians were there since every time a gun fired the tiny room filled with soot and had to be cleaned.

Like his contemporary Sam Hurry, Brian developed a fascination in the soldiers activities and was often to be seen in their company and riding on their Scammel gun tractor as they rode to the common pump in Walsingham to fill their water containers. That fascination was to end after eighteen months when, as the threat of sea-borne invasion reduced, the Newfoundlanders and their heavy artillery moved out. At first the site was used as a decoy for an airfield at Docking a few miles away but soon it was announced that an operational airfield was to be built there.

Farmland and a small road were soon acquired and construction work commenced when Taylor Woodrow moved onto the site. All this gave Brian an extra reason to hurry home from school since there was the chance of a ride on earth moving machinery. Five or six yellow tractors hauled levellers and graders to iron out bumps and minor gradients. When their work was completed, bulldozers moved in flattening hedges and creating large spoil heaps. In an area where horses were still in use on farms, the chance to watch and ride massive machines was an experience not to be missed.

When the site was cleared, work on laying the runways started. Bags of cement were delivered by coasters and unloaded at Wells quayside, continuing their journey to the construction site by road. Barrage balloons were tethered to each ship as it unloaded to deter any would-be air attackers.

By now the war was beginning to have an effect on Brian's life. In order for the airfield to be built the cottage where he lived had to be demolished and his family, and another from the adjoining cottage, were rehoused in a large farmhouse nearby. This had been partitioned off to preserve a semblance of privacy for both families and part of the furniture in each 'half' included a Morrison table shelter for use in the event of air raids.

The quiet world around Egmere was being changed, quickly and surely. One night as Brian played indoors he heard sharp cracking sounds and then silence as work on building stopped. He discovered next day that a German aircraft had tried to destroy the large cement mixer that was in use day and night. The incident was enough to scare two families who were living in temporary caravans near the farmhouse while their menfolk worked on the site. Expecting more bombs to drop on successive nights they quickly moved out, although the bombers did not return.

Suddenly he was aware the world at war was a dangerous place, and with the creation of an airfield in the next field, that world was now on his doorstep. He was too young to understand it fully or even begin to make sense of events. All he knew was an overpowering sense of fear, wondering just what would happen next. A local landowner was kind and reassuring and he believed her when she said that everything would be alright. From that point bravery would be the order of the day and he would try and continue as he had before without worrying too much.

With this new found bravery he found himself climbing on the rubble of his former family home to gain a better view of the Halifax and Stirling bombers taking off. Then, in the early hours of morning he would lay awake and hear them return, circling and waiting permission to land. There was the possibility of extra pocket money to be earned for the station's commanding officer was billeted with the family in another part of the farmhouse. He regularly gave Brian extra pennies for keeping quiet while he caught up with sleep after night duties. He even became the source of special passes, enabling the youngster to attend ENSA concerts and film shows on the airfield.

While all this went on in Norfolk, at an Operational Training Unit "somewhere in England" a new recruit to the Royal Air Force was completing his training and awaiting details of his first posting. Albert Berry left his job as a clerk in the City of London when called up for military service on September 15, 1942. Attendance at evening institutes after leaving school left him qualified to be considered for pilot training. His eyesight had failed the strict medical but he had successfully completed studies in wireless operation and air gunnery and was destined to realise his ambition to become aircrew. His posting was to 199 Squadron based at a newly built airfield in the heart of northern Norfolk: Egmere airfield had become RAF Station North Creake, the postal address Wells next the Sea, and it was here the young NCO was to start his service career. Within days of arrival at North Creake he was posted to 171 Squadron based at the station.

Eighty miles to the south in an aircraft factory in Hertfordshire a young Norfolk girl was becoming homesick and frightened by regular air raids on London and the Home Counties. For Jean Kershaw it was a massive culture shock after her previous work in service to a local family and later driving a mobile canteen for a firm building an airfield near her home village in the early days of war. With this last experience to draw upon, she applied to the NAAFI and was successful in being appointed to a new job as charge hand on the sole mobile canteen based at RAF North Creake.

The work of the squadrons based there was in the field of radio and radar jamming. For 199 Squadron that meant aircraft equipped with Mandrel equipment to jam enemy radar and flying ahead of the main attack with jamming devices operating to confuse the enemy. Planes from 171 Squadron dropped Window, metallised paper strips which would suggest that one solitary plane was part of a large force. Window registered many times on enemy radar screens, confusing observers into believing a large attack was about to take place. As war progressed both squadrons would carry bombs in addition to their regular duties in radio counter measures.

Accommodation for the young crewman was in a Nissen hut, beds arranged down each side, a small locker beside each one and a tortoise stove in the middle. The crew room was brick built and here wireless operators, gunners, pilots, navigators and flight engineers would gather each day to receive their instructions. The crew room was part of the flight section building and on one of the walls was a picture of a Halifax bomber. Other section buildings and messes had their own decorations on walls, but in one of the station buildings the portrayal of a Short Stirling had a particular tinge of sadness surrounding it. That particular aircraft had been one of a force of 15 which took off from North Creake in June 1944, just three months before Albert Berry's arrival there. Fourteen aircraft returned, the plane whose image graced a hutment wall was lost without trace One hundred and seven sorties were flown from Creake that month and it was the only loss. In memory of the eight crew who had lost their lives the letters RIP had been added to a corner of the picture.

Home for Jean Kershaw was a flatroofed maycrete near the Institute. She would rise early each morning, eat in the NAAFI staff dining room and prepare for her round. The van would be loaded with cakes, buns, sweets and cigarettes and the boiler switched on to keep hot a weak brown liquid that passed for tea or coffee. After cutting slab cake into profitable portions she commenced her tour of the aerodrome stopping at prearranged points for up to fifteen minutes. When the round was complete she drove back to the Institute to restock, clean out the urns, and repeat the trip in the afternoon. Leisure time she spent by choice in her billet reading, knitting or talking to other NAAFI girls. Females were in short supply on any airfield and in great demand. Jean had many offers of dates but chose her suitors carefully, feeling that some personnel didn't seek her solely for her personality!

Leisure time for Albert Berry was spent with his crew in games of billiards, darts or snooker in the mess, or at the station cinema. Occasionally it would be his turn to be station orderly officer and he would don the black armband of authority.

Outside normal hours there was a rota officer on duty whose job was to deal with any complaints or non military problems which might arise. Most often the problems were complaints about food – there were plenty forthcoming from the airmens mess. Traditional rivalry between messes had left the airmen feeling that food in the sergeants mess was of higher quality. It was the orderly officer's job to make sure complaints were heard by the cookhouse sergeant. At other times he would be called upon to liaise with relevant departments on the station to ensure that a leave pass was amended so a WAAF girl could catch an earlier train to begin her leave.

* * * *

Passes and pieces of paper were vital to the smooth running of an airfield but they didn't get in the way of Brian Beaumont enjoying himself. On occasions he forgot the pass allowing him entry to the 'drome but he still managed to convince the orderly officer he had a right to be there. Instead it meant he had a seat at the back of the cinema instead of the front where he would be conspicuous and likely to be challenged to show his pass.

By now he was well used to the airfield. Near the farmhouse which was his temporary wartime home was a dispersal point which, as well as being home to aircraft, was the billet for Wingco, the affectionately named billy goat who was a squadron mascot. When not smuggled aloft by crewman, Wingco took great delight in leaving the dispersal point and making for the farmhouse where he would devour every flower in sight, then appear at the french windows, mouth stuffed with flower heads, demanding more to eat.

Road tankers carrying aviation spirit were a regular sight on the dry road, part of which had now been closed to the public because of the close proximity of road to the airfield technical site. Thus the journey time to Wells or Fakenham was almost doubled as locals were forced to use byroads to circle the airfield. Sometimes the rural carriageways were insufficient to accommodate extra traffic generated by the airfield. One day on his way to school Brian's bus became wedged against a RAF truck travelling in the opposite direction. There were only a few inches clearance either side and neither driver was willing to try passing. The vehicles remained where they were, stuck tight between the bank, the hedge and each other. The only way Brian could continue his journey to school was on foot but first he had to leave the bus. Since the side doors could not be opened he and his friends piled through the emergency exit at the rear. The air that day was blue, he remembers, and it was not the colour of the uniforms!

During summer holidays when days were long he worked in fields beside the airfield. He watched as American Flying Fortresses made emergency landings, too badly shot up to be able to continue to their home base. He saw crew members climb down from the aircraft, dazed or injured. He watched trucks drive close to the planes where men standing on the platform at the rear lifted bodies of dead crewmen from inside the fuselage.

* * * *

Sunday nights were open nights in the sergeants mess at North Creake, when WAAFs and NAAFI personnel were invited along. For Jean Kershaw open evenings gave her the chance to meet male members of the squadrons in relative safety and she soon found a young wireless operator occupying a place in her thoughts each day. So taken was she by young Albert Berry that she invited him to her twenty-first birthday party held in the NAAFI. From then on romance blossomed and the mobile canteen which was normally forbidden to sell to anyone above the rank of sergeant and intended primarily for groundcrew unable to get back to their canteen for a break, soon began to stop to order when hailed by the NCOs of a certain aircrew.

Not that other personnel minded much. The local girl had already endeared herself to many on the station by her frequent trips to the NAAFI warehouse in King's Lynn. Very rarely did the trusty Ford travel empty to collect supplies. Although only the driver and mate were visible in the cab, the servery area at the rear was often full of men keen to reach the railway station at Lynn which boasted more frequent departure times and better connections than the branch line terminus at Wells. On one occasion the vehicle overheated from the extra load as it laboured up one of the county's rare inclines. While men from the MT section travelled out to repair the beleaguered van the illegal servicemen passengers continued their journey on foot while the driver and mate put up for the night at the local YWCA.

With a boyfriend in aircrew there was a different pattern to Jean's evenings. After work in the kitchens and a few hours in the billet she would cycle out to the perimeter track to watch the planes take off. On their return in the early hours of the morning the crew would always try and fly low over the hut, a prearranged signal that all was well. For airmen there were seven days leave every eight weeks. Sometimes the NAAFI girl was able to arrange her leaves to coincide and there would be time spent visiting families: a train ride to the East End of London or a cycle excursion along the lanes to her home a few miles away.

* * * *

The Christmas of 1944 was shared together on the airfield. A full itinerary was arranged to suit the season and a printed programme distributed to all personnel carried the paragraph: "Although the arrangements detailed in this programme must of course be subject to operational commitments it is hoped that in any event the major portion of the arrangements will be carried through." There followed a cautionary note about Christmas post which because of "the considerable delay in the receipt of Christmas Mail and the heavy commitments of GPO staff throughout the country, an assurance is given that all mail received in the Unit will be distributed immediately on receipt." To expedite the process the camp post office would be open for two and a half hours on Christmas Day and two further hours on Boxing Day.

The festive programme began on December 23, a Saturday, with a football match against RAF Docking. In the early evening 'Liberty Runs' using station transport went to Lynn and Fakenham, leaving those towns at half past midnight for the journey home. For those who could resist the lure of the town there was a performance of Candlelight staged by ENSA in the Station Cinema.

Christmas Eve opened with a church service – Anglicans on the station worshipped in the Station Church, the other majority group there met together in the R C Hut! In the afternoon there were bicycle races around the perimeter track with entrants expected to use Service bikes and pay an entry fee of 6d (2½p) each. Fees collected would be distributed as cash prizes to the first 10 home in the final, taking place on Christmas Day. A barrel of beer would be awarded to the Unit with the most personnel among the 10 winners and there would be a tote competition with another barrel as prize. The evening saw more Liberty Runs for those unable or unwilling to spend an evening in the Camp Cinema watching Cary Grant in Once Upon A Time, closely followed by the station concert party giving a recital. Later in the evening the Station Glee Singers made a tour of nearby Holkham, Wells and Walsingham, singing carols and raising money for Wells cottage hospital. They were back in time to lead community singing at the dance, by now well under way and open to all personnel below the rank of corporal.

Christmas Day heard the tannoys crackling with festive music, rallying the tired and befuddled to breakfast. Christmas Communion, station parade for carols and a church service for Catholics took place before a comic football match with officers versus the sergeants. The grand final of the cycle derby was run, followed by a cycle race over a shorter course for WAAF personnel where the barrel of beer had been replaced by a cash prize for the first three racers past the winning post. At mid-day the station commander visited sick-quarters before setting out on a tour of the messes to take his Christmas greetings. The festive dinner took place in four sittings and lasted throughout the afternoon. Afterwards, football, tug of war contests and films were available to lead up to tea, served as a cold buffet. In the evening a party open to all ranks and relatives was held. "Santa Claus will distribute everything but a whisky and soda," said the programme. "Spot prizes, novelties and free refreshments (except beer!)."

To bring their guests to the party, personnel were able to arrange station transport to tour the surrounding towns and villages. Names were delivered to the Guardroom by the appointed hour and transport rotas drawn up – the collection details being broadcast on the Tannoy well in advance. No charge was made for the journeys but all goodbyes would have to be said on the airfield as personnel were not allowed to travel on the vehicles taking or returning guests.

Boxing Day must have dawned wearily for those still fresh from the party. There was another football match for those with the enthusiasm and energy to spare but most preferred to idle away the hours before a dinner of Roast Duck and green peas. Then, with energy levels replenished there was a chance to play rugby – "Officers versus the Rest." For the less energetic the Station Cinema had changed its programme and there was yet another Liberty Run in the evening

* * * *

Seven months into 1945 the fortunes of war were changing and Albert Berry's squadron was disbanded. Transferred to 192 Squadron he was posted to Foulsham nearby and his association with North Creake ended. 199 Squadron was disbanded a few days later. Jean Kershaw was sent to RAF Bircham Newton with its fine mock Georgian architecture and placed in charge of a Morris overseas version of a mobile canteen. But before the Squadrons left North Creake there was one last event and an open day was held. The Bishop of Norwich attended and in a hangar, from a pulpit set in front of an aircraft, a service of thanksgiving was conducted.

Brian Beaumont remembers the service and the day when the Tannoys heralded the end of war and the commanding officer's voice echoed around quiet fields as he thanked his men for their work. Later on Victory in Europe day, there would be a day's holiday from school.

With hostilities ended life on the airfield changed. There was still occasional flying but nowhere near as busy as before. Mosquito bombers flew into North Creake, in use now as a store and eventually a breaking station for aircraft excess to requirements. The airfield finally closed in 1947 and six years after he had watched earthmovers level the site, Brian Beaumont witnessed tractors plough the grass around the runways and restore it back to agricultural use.

The watch tower, hangars and some of the Nissen huts around the technical area remained intact and were taken over by the firm of British Field Products. Among buildings which survived as a store was one with a mural painted on its end wall, a mural bearing the legend RIP and depicting a Stirling bomber. Spotted by aircraft enthusiasts the wall was dismantled and the mural removed intact for transporting to the Bomber Command Museum at Hendon. It had been painted by Flt Sgt Ted Allen while he was stationed at Creake and unveiled in 1984 by his widow. Her late husband had often shown her paintings he had completed on walls at the airfield but never this particular one. "They never spoke about planes that were lost so I never knew about it," she told pressmen.

The leftover buildings, now used by an agricultural feedstuff company, are a regular sight to Brian Beaumont who continues to live in the area. He eventually married a girl whose family lived on the other side of the airfield and whose house had also been a billet to aircrew. In the house he now shares with his wife hangs a martingale with four horse brasses. Part of one of the brasses had worn and the family had asked their serviceman lodger if he could get it repaired on the airfield. The job was completed, and a craftsman's job done, but not before the martingale and its brasses had been taken over Germany on a bombing raid by a crew keen for a good luck charm.

Jean Kershaw and Albert Berry continued their romance and when war was over they married. After his demobilisation he and his new wife eventually settled in Fakenham, destination of all those wartime Liberty runs. Twenty years later, as was common practice then with driving tuition, he was taken to gain experience of handling a car on the perimeter track of an old airfield. Following the driving instructors directions he found himself taking the dry road to Wells and halting at North Creake and the remains of the old Egmere airfield. There three point turns in an Austin A40 took the place of bumps and circuits in a four engined Stirling bomber.

Much of the technical site at Egmere airfield (later RAF North Creake) was intact when Jean and Bert Berry revisited in 1988. Beyond the hutments in the background stands the flat roofed control tower, two chimneys in place where the wooden 'watch office' perched. (Author)

***Below** Airfield activity 1980s style: a mobile crushing plant and heavy dumper trucks turn a perimeter track into a mere pile of dust and rubble. In 1988 the last traces of the runways at Great Massingham airfield were being wiped from the face of the earth.* (Author)

17 Aftermath

Life is never the same after war. Social upheavals are inevitable and go hand in hand with conflicts and when peace is eventually won the old order never returns. Indeed, in the wake of World War Two there was a social revolution with the birth of the welfare state and the beginnings of an education system which would last well into the 1980s. In the years of peace, the Royal Air Force was not spared its share of change and upheaval. Its operational strength was significantly reduced by demobilisation when those drafted for hostilities-only returned to civilian life. Its peace time role no longer required vast numbers of airfields, there were not the personnel to staff them and strategically many of them were excess to requirements anyway.

Some, like North Creake, were used to store planes and supplies from 'dromes in advanced stages of closure and they became the focus for regular sales of surplus equipment. Others took on a new lease of life with the introduction of National Service in the 1950s. Many were put into 'mothballs' prior to closure at some later date. Part of the revolution affecting the RAF was concerned with modern technology and new defence strategies. The development of jet engined planes and missile defence systems meant the need was for fewer airfields, and those in use would be more streamlined than their wartime precursors. Gradually many of the old wartime sites were sold off, many to their original owners. Sales of former airfields continued through much of the fifties and sixties and the reduction was hard to ignore. During the heyday of the airfields there were 35 operational stations in Norfolk equipped for planes to land and take off. Now regular military flying takes place at just two, with an additional site continuing to be maintained for 'standby' purposes. The airfields which have survived into the latter half of the century have changed beyond all recognition. Indeed, the word 'airfield' conjuring pictures of the grass strip at Bodney with timber huts either side no longer seems to describe accurately the modern sprawling installations where multimillion pound aircraft fly and advanced surveillance systems monitor skies and perimeter fences alike.

The fortunes of old airfields were many and varied. Before total closure Debach in Suffolk became a holding camp for German prisoners of war returning to their homelands. Wymeswold in Leicestershire has now reverted back to farming use but its former satellite at Castle Donington is covered in glory as the East Midlands Airport, acting as starting point for many package holidays. Horsham St Faiths airfield is now home to Norwich airport, with some of its buildings in use as an industrial estate. At Snetterton part of the site is in use as a motor racing circuit and one of the original hangars there was recently turned into an ice skating rink by a television company filming a lavish production of The Sleeping Beauty on ice.

The former home of American B17 Flying Fortresses was chosen as being the only suitable building large enough to house the vast array of sets, lights and cameras.

The airfield at Methwold in Norfolk was quickly run down and its buildings emptied. Then, as now, housing in rural areas was scarce and while there was some property available for renting it was often beyond the means of many newlyweds. Peggy Newton found herself in this position needing accommodation and was quick to realise the potential of airfield huts going to waste. She visited the commanding officer at nearby Feltwell to discuss moving into the empty rooms. She remembers her meeting with him. He could not encourage her to move in, yet hearing of her plight he did not forbid her to do so. Accordingly Peggy and her husband moved in to the deserted buildings, the first of many to do so long before the word squatter became fashionable. She stayed for nine months in the former airfield hospital and it was appropriate that her first child should be born there.

When the floods of 1947 ravaged much of East Anglia the airfield buildings at Weeting were full to capacity with families unable to live in their own homes because of damage. Occasionally the village policeman came visiting but he turned a blind eye to the airfield's new occupants. "I'm supposed to tell you you're not meant to be here," he said one day before cycling away.

It was her first home and she retains fond memories of it, still enthusing about the lino and the heat the tortoise stove generated. "You could hear the rats running over the hut but that didn't seem to matter," she says now. Eventually she moved back to her parent's home for a short time after her brother moved out creating more space. She still lives close to the old airfield, which survives as a hangar used by an agricultural firm and a few ruined brick hutments in trees beside the main Stoke Ferry to Brandon road.

* * * *

When war ended Great Massingham reverted to its original purpose as a satellite for nearby West Raynham before being sold off in the late 1950s. As this book was being written the last traces of perimeter track were being removed and there is now little to remind that an airfield had once been there. For Jim Paxton who served there, service life continued even when hostilities had ended. In order to get away from home he signed up at the age of 15 years one month and ten days on a twelve year engagement. Since service could only be calculated from the eighteenth birthday his contract was effectively for fifteen years. When he eventually returned to civilian life he worked in the rubber and nuclear industries. Premature retirement followed in the early 1980s and he now lives back in his native Scotland.

He remembers airfield life with affection and tells of bonds of friendship forged then which have lasted a lifetime. But it was not without personal cost. Like many servicemen, he still feels he was robbed of an important part of his youth, denied the chance to enjoy to the full his teenage years.

"On return to civilian life I was lucky to get the jobs I did. I fared better than some of my contemporaries but in the race of life I felt I had to run hard just to try to begin to catch up with my classmates from schooldays. I had a handicap of fifteen years in that race," he says today.

In Yorkshire the former hutments of Pocklington airfield are in use providing accommodation for firms on a light industrial estate. Much of the old site has been returned to agriculture but a grassed strip remains and private aircraft land and take off there. Holiday traffic heading from the north and Yorkshire dales to the ferry port at Hull jams the former Roman road skirting the airfield and modern descendants of the Pocklington villagers who complained at the noise of night flying still have their lives disturbed by speeding lorries and cars.

Edith Kup was demobilised in December 1945 and it took her a long time to settle back into civilian life after the communal experience of airfields which did much to shape her independent spirit. "I found I no longer spoke the same language as my family and, though I understood why, I very much resented being asked what time I was to come in. Civilian life seemed light-weight and values were very different," she recalls.

She moved to Scotland, married, divorced and raised her two daughters. When they had completed school she went to Germany with the Women's Royal Voluntary Service working in the mess at RAF Rheindahlen. She reverted to life on a high-tech airfield in the front line of the current defence plan as if she had never been away from it. After two and a half years she returned to the UK, joining Kent County Council as a social worker. Now retired, she has returned to her native Yorkshire.

Remembering her airfield days she says: "I learned a lot about people which has stood me in good stead ever since. Our general conduct outside the airfield was very light hearted. Our business was grim in the extreme – we never knew who would be there tomorrow. We all supported the crews to the best of our ability and we learnt the hard way about death.

"It also made me appreciate things like the countryside and the forces of nature. I think the effect of it all was to teach me what really mattered. It gave me a strong sense of faith and now I can't understand anyone who can live life without it."

Unlike many of the airfields mentioned in these pages, Honington is destined to continue into the 1990s. When the Americans moved out in 1946 the station operated under Transport Command who carried out runway improvements there. During the 1950s and 1960s there were periods of temporary closure but the station burst into life in the 1970s with the arrival of Buccaneer jets. In the early 1980s they were replaced by Tornado multi-role combat aircraft whose modern bomb proof blister shelters contrast strongly with the large wartime hangars still on the site.

John Wortley has remained rooted to that part of Suffolk, continuing to live within sight and sound of Honington. As a county councillor he is heavily committed to serving the community and is quick to highlight the advantages available to local people with a modern airfield operating beside them.

Benefits were felt during those early days of construction when a mains power line had to be redirected from its previous course which crossed the camp. Its new route brought it nearer to Fakenham Magna and as a result the village was connected to mains electric power years earlier than had been anticipated. Later, hangar construction at the base, much of it within the parish boundaries, has given financial benefits. Local civilians are employed and businesses still trade with Honington. But there is one drawback he sees in the 8 am and 5 pm 'rush hour' when personnel are heading to and from the station and crowding the narrow twisting roads. Not a good time to be out for a leisurely walk he muses.

"I like the noise of our planes," he says. "During 1939–1945 we heard and saw German planes. They were not a nice sight since their object was to damage and destroy. For our protection we should never complain about red, white and blue flying over, although the value of our property is very much reduced."

The contact which young Sam Hurry had with American groundcrew on dispersals near his home made a lasting impression on him. After much of his working life spent managing a shoe shop he now lectures in retailing at a Norfolk further education college. All his spare time is devoted to helping create a memorial museum marking the 100 Bomb Group's time in East Anglia. Although many of Thorpe Abbotts runways and buildings have been lost, two Nissen huts and the control tower have been restored by volunteers to their wartime appearance and now house relics and reconstructions from the American occupation there. Many of the exhibits on display have been sent over by American veterans who were stationed at the airfield.

Forty-five years later Sam Hurry regularly makes a 120 mile round trip returning to the place of his boyhood to work on the museum. Weekdays will see much of his spare time occupied administering the project and writing letters maintaining the vital links between the enthusiasts and volunteers in the United Kingdom and veterans back home in the States. Those few years in the shadow of an operational airfield left a deep and lasting impression of the former schoolboy and the friendships made then have led to a lasting association with many of the servicemen.

When Colin Acock's work transporting ballast to airfield construction sites and driving mobile canteens for the NAAFI came to an end he was sent to Pulham Market. This disused World War One airfield still had its airship hangars intact and they were being used by the RAF as a stores recovery unit. As a civilian driver employed by the Air Ministry his work involved ferrying defused bombs to local railway stations ready for transporting to Wales where their contents would be steamed out. After five years the work came to an end but he continued as a driver, working for grocery store chains.

Recalling his service on wartime airfields he is left with an over-riding feeling that despite the hardship encountered by all personnel living and working on them, there was a greater value placed on life. "Money wasn't the root of all evil as it is today. We worked hard and in conditions of danger sometimes yet we never went on strike. Looking back I don't think anyone was really happy at living in close proximity to an airfield but they did provide a good deal of work," he says.

Of all the airfields served by Mary Williamson in her NAAFI mobile canteen only two have runways still intact. University air cadets fly from West Raynham whose main use today is as a missile base. Sculthorpe continues as an American standby base and although no squadrons are based there it regularly plays host to exercises and has been a temporary home to planes from Lakenheath when runways there were being resurfaced.

Mary is now in her eighties and spends her time making and selling soft toys to raise money for local charitable causes – including the cottage hospital which the Glee Singers from RAF North Creake once adopted as a cause. A few fields away from her neat bungalow is the parish of Little Snoring where the airfield is now home to a local flying group. The control tower, a runway and a hangar remain and small aircraft still echo in the skies above her home, although the sound of the single engined bi-planes is more suggestive of the earliest days of flying.

* * * *

 Where airfields have survived they have been put to use in ways more diverse than anyone could have thought of when the decision was made to sell them. Turkey farms, industrial estates, vehicle testing tracks and prisons have all been developed on the remains of wartime airfields. Inevitably though, in some places the past still holds sway and tales of ghosts and time warps abound. In Lincolnshire a dark ghostly figure has been observed standing near the control tower moving his arms as if waving to planes taking off. Sam Hurry tells the story of dogs unwilling to enter buildings on the former domestic site at Thorpe Abbotts. He tells of a colleague working on the project to convert the tower into a museum experiencing a sudden chill throughout the building, a wave of cold engulfing every room and bringing with it the noise of aircraft and the bustle of the tower as it once was. Now before work is carried out in each room the workers whisper to the silence just what they will be doing and the spirits of Thorpe Abbotts rest easy.

Most spectacular and well documented of all ghostly happenings on old airfields takes place at Bircham Newton. Much of the airfield domestic and technical buildings remain intact and in use as the Construction Industry's Training College. Young apprentices are now boarded out in the former billets and their evening conversations after a day in the classrooms (former messes) or building brick walls in the old hangars are probably far removed in subject from the crew of an aircraft who once discussed the afterlife. Those young airmen made a pledge that if anything ever happened to them they would return and meet up at the station squash courts where they had spent so much of their off duty times. A few days after that discussion their aircraft crashed near the airfield, killing them all.

Since then the spectre of a figure dressed in RAF uniform has appeared in the gallery of the squash court. On one occasion the vision was witnessed by a member of a film crew working at the site on a training film. He returned later with a colleague and a tape recorder, leaving the latter in the squash courts switched on to record. They came back next morning and played back the tape, discovering that it was full with the noise of aircraft and machinery.

The tape was subsequently broadcast by the BBC who received letters from listeners complaining that their pets had been disturbed by the noises. Intrigued by this the BBC decided to investigate further and sent two spiritualists to the airfield. On entering the squash courts one of the psychics went into a trance and began speaking in another voice, a voice which told the story of the airmen's meeting and their pledge to return to the airfield squash courts if anything happened to end their lives.

* * * *

This book came about because of a wish to know more of ordinary life on and around airfields. It has not been the intention to create a picture of them peopled by ghosts and phantoms. By sharing the experiences of people who were there something of the uniqueness of wartime airfields can be shared. They were built at a time when aviation was still in its infancy and the tactics of airwar were still being evolved as battles and bombing offensives took place. In that short time airfields grew from being grass strips with tents and barns housing men and machines, to communities where as many as two thousand people lived and laboured to keep two hundred aircrew airborne. They became places not just to work, but to live and play and in meeting all those diverse needs they grew into small towns with their own electrical and water supplies, their own hospitals, cinemas and churches. They developed in areas of the countryside chosen not for their communications or facilities, but simply because the land was flat and within easy flying distance of Europe. Quiet villages and hamlets suddenly found themselves with a settlement down the road made up of hundreds of young people from all over the country, all over the Empire. They had been taken from ordinary, often mundane jobs, trained and then sent into remote countryside to play their part in winning a war. For service personnel and locals alike it was a culture shock.

There had been nothing like it before and, with the exception perhaps of constructing an airfield on the Falkland Islands in the 1980s, there has been nothing on such a scale since. Now that developments in armaments and new defence initiatives have advanced it is unlikely that there will ever be the same need for airfields in the future. The years from 1939 to 1945 were an important time in England's history and airfields were a unique part of it; they were built for a particular war and when they had contributed to its victory there was no longer need for many of them. A few continued in operational service and became home to a generation of aircraft bearing little resemblance to the lumbering, piston engined bombers which once flew from them. Those have continued to be home and workplace to a new generation of people who have chosen the Royal Air Force as a career. Although their fundamental task remains unchanged, their daily life styles living and working on an airfield are many years removed from those recorded here. Yet in these days of defence budget cutbacks their legacy remains in the old buildings left over from a once in a lifetime period of expansion and growth. And somewhere it remains rooted in the hearts and memories of people who, half a century ago, were part of the heyday of airfields.